Descartes' Temporal Dualism

Descartes' Temporal Dualism

Rebecca Lloyd Waller

LEXINGTON BOOKS
Lanham • Boulder • New York • London

Published by Lexington Books
An imprint of The Rowman & Littlefield Publishing Group, Inc.
4501 Forbes Boulevard, Suite 200, Lanham, Maryland 20706
www.rowman.com

16 Carlisle Street, London W1D 3BT, United Kingdom

British Library Cataloguing in Publication Information Available

Library of Congress Cataloging-in-Publication Data

Waller, Rebecca Lloyd.
Descartes' temporal dualism / Rebecca Lloyd Waller.
pages cm
Includes bibliographical references and index.
ISBN 978-0-7391-7522-4 (cloth : alk. paper) -- ISBN 978-0-7391-7523-1 (electronic)
1. Descartes, Rene, 1596-1650. 2. Time. I. Title.
B1878.S67W35 2014
115.092--dc23
2014034965

∞ ™ The paper used in this publication meets the minimum requirements of American National Standard for Information Sciences Permanence of Paper for Printed Library Materials, ANSI/NISO Z39.48-1992.

Printed in the United States of America

Contents

Acknowledgments

I would like to thank a few of the people who have supported me in my scholarship—in particular, those who generously offered their input and critiques on some of the ideas in this book. I would first like to thank Jan Cover, Michael Jacovides, Jeffrey Brower, and Martin Curd for the support and criticisms they offered while I was studying at Purdue. Thanks also to Kimberly Blessing, whose long-standing support has left an indelible mark on all of my philosophical work. And, finally, I give my heartfelt thanks to my husband, Jason Waller, whose generosity, philosophical acumen, and intellectual honesty improves all my work (and all of my life) in more ways than I can express.

NOTE ON ABBREVIATIONS

AT: This abbreviation refers to the original language text of Descartes' works. The full citation is: Descartes, Rene. *Oeuvres de Descartes*. Edited by Charles Adam and Paul Tannery. Paris: J. Vrin, 1905. The 'AT' will primarily be used in this text to indicate the line numbers cited in the English translation by Cottingham, Stoothoff, and Murdoch. The 'AT' abbreviation is followed by a volume and page number.

CSM: This abbreviation refers to the English translation, whose full citation is: Descartes, Rene. *The Philosophical Writings of Descartes Vols 1-2*. Edited and Translated by John Cottingham, Robert Stoothoff, and Dugald Murdoch. Cambridge: Cambridge University Press, 1984. Reprinted with the permission of Cambridge University Press. The 'CSM' abbreviation is always followed by a volume and page number.

CSMK: This abbreviation refers to the English translation, whose full citation is: Descartes, Rene. *The Philosophical Writings of Descartes: The Correspondence*. Edited and Translated by John Cottingham, Robert Stoothoff, Dugald Murdoch, and Anthony Kenny. Cambridge: Cambridge University Press, 1991. Reprinted with the permission of Cambridge

University Press. The 'CSMK' abbreviation is always followed by a volume and page number.

Introduction

In studying Descartes' views on time, the challenges are immediately apparent. One quickly discovers that time plays various important roles in Descartes' physics and metaphysics, and likewise finds that understanding these roles is extremely difficult. Understanding these roles is difficult because Descartes' own texts offer so few discussions on time, and there are numerous, long-standing disagreements concerning how to interpret the few that there are.[1] Since Descartes' comments on time are so limited that even learned commentators have failed to reach anything near consensus, it is natural to feel hesitant to enter the fray and propose a "new" interpretation of Descartes' account. It is tempting to ignore Descartes' account of time, under the assumption that his views were either too poorly defined, or too poorly expressed to offer valuable material for study. Yet, when one recalls the various important roles that Descartes gives to time, it seems that one cannot ignore Descartes' account without leaving an important gap in Cartesian scholarship. Thus (showing either bravery or foolhardiness), I offer this book into the fray.

The important roles that Descartes gives to time are not hard to recognize. In *Principles II.23*, for example, Descartes indicates that time is necessary for distinguishing bodies since he claims that "any variation in matter or diversity in its many forms depends on motion."[2] In making diversity dependent on motion, Descartes makes it equally dependent on time, since he claims that there can be no motion without time.[3] Descartes also assigns time the important function of grounding the Second Causal Argument in *Med III* when he claims that one need only consider "the nature of time" to see that God must exist.[4] Given the important roles that Descartes gives to time in both his metaphysics and physics, it seems appropriate to wonder why a thorough account on the nature of time itself appears to be missing from the Cartesian corpus. In this book, I propose a rather simple answer to this question; namely, that Descartes' account was presupposing the accounts of time that were generally known and followed by Descartes' predecessors and contemporaries.

It is certainly true that Descartes offers very few discussions on the nature of time. Likewise, it is true that these sketches, when considered in isolation of their historical context, offer frustratingly few details upon which to ground a Cartesian account of time. When considered from within this context, however, the texts seem far more fruitful. Indeed, I believe that much current confusion concerning Descartes' views has resulted from interpreters attempting to understand his account without first considering the alternative approaches available in the seventeenth century. If Descartes' account is considered in a historical way then his view sounds particularly odd. Indeed, contemporary ways of conceiving of time are so foreign to the ways in which time was conceived by Descartes and others of his era, that today's "basic" understanding of the subject (which contains a background that includes Newton, Einstein, McTaggart, etc.) is a hurdle that must be overcome if Descartes' account has any hope of being understood.

Thus, in this book, I attempt to interpret Descartes' account of time by reading his texts in light of the issues and disagreements that concerned Descartes' predecessors and contemporaries. I thus offer a historically sensitive approach to interpreting Descartes' views on time, which attempts to expand upon Descartes' otherwise brief (and apparently inconsistent) discussions. When viewed from within this context, Descartes' discussions are not only significantly clarified, but they can also be used to ground a new interpretation of Descartes' account of time. As one might expect, however, the background assumptions grounding Descartes' account are so radically different from today's background assumptions that the new interpretation I offer will sound pretty odd.

Over the next several chapters, I will argue for the view that Descartes maintained a type of temporal dualism that had certain characteristics in common with his historical contemporaries, but that nonetheless maintained a uniquely Cartesian character. Of note in this account, will be the claim that Descartes maintained two temporal attributes: (1) successive duration and (2) an innate idea of time. The latter attribute, which I will call time-in-thought, is the more radical claim, but one whose significance will be explained and defended toward the end of the book. Indeed, I will argue that this latter temporal attribute has been overlooked in a way that has led to much of the confusion that has enveloped contemporary discussions on Descartes' account of time. By failing to give sufficient notice to the nature and role of this second temporal attribute, commentators have missed a key feature of Descartes' account which is capable of resolving a number of puzzling problems in the scholarship. When the innate idea of time-in-thought is brought to the forefront, one sees that Descartes' account is less obscure than it may otherwise seem and that it poses an important development in the historical conceiving of time. Indeed, his account might even be read as suggesting an interesting precursor to Kant's *Transcendental Aesthetic*.[5]

This first book-length study of Descartes' account of time will compose five brief chapters. The structure is as follows:

CHAPTER ONE: PRE-CARTESIAN APPROACHES TO TIME

I begin the book by providing a historical context for a study of Descartes' philosophy of time. I do this by outlining the alternative views on time that were prevalent among Descartes' contemporaries. I argue that a basic familiarity with these views brings clarity to Descartes' account, since during the seventeenth century, all accounts of time were structured around these various alternatives. Analyzing these views reveals the three most important questions to ask of Descartes' account. These questions are:

1. Is time an independent or dependent entity?
2. Is time dependent on motion, or in what way is it dependent on motion?
3. Is time a mental entity?

Having argued for the centrality of these three questions, I then indicate how Descartes answers each of these questions in the three following chapters. I deal with the dependence question in chapter 2, the question of time's relationship to motion in chapter 3, and the question of time's relationship to minds in chapter 4.

CHAPTER TWO: WHY CARTESIAN TIME IS NOT (AND COULD NOT BE) A SUBSTANCE

In chapter 2, I argue that time cannot exist as an independent entity for Descartes. Insofar as independent entities are substances in Cartesian metaphysics, I argue that time could not be a type of Cartesian substance because (i) there would be no grounds for a substantial time to be divisible into the numerically distinct parts/moments that Descartes' account requires and because (ii) a substantial time would contradict the *Med III* claim that any substance can exist at one isolated moment. Since Descartes' ontology exhaustively divides between substances and attributes (i.e., that which depends on substances), I thereby conclude that time must be a type of attribute for Descartes. As this fact is explicitly stated in *Principles I.57*, this chapter provides both an explanation for *I.57*'s assertion and grounds for thinking that *I.57* offers a significant statement on the temporal account that Descartes consistently maintained.

CHAPTER THREE: DESCARTES' TEMPORAL DUALISM

In chapter 3, I address time's relationship to motion via a close analysis of *Principles I.57*. By considering the relationships between motion, duration, and time, I demonstrate that Descartes' account of time includes two really distinct temporal attributes (namely, duration and time-in-thought). These two attributes are really distinct since time-in-thought (the mode of thought by which duration is measured) is dependent on motion but motion is not directly dependent on time-in-thought. In contrast, duration (the successive enduring of any thinking or extended substance) is not dependent on motion but motion *is* directly dependent on it. From these differences, I establish that these two attributes are really distinct and also that they appear to be very similar to the "extrinsic" and "intrinsic" times found in Suarez.

CHAPTER FOUR: TWO TEMPORAL ATTRIBUTES THAT ARE ONTOLOGICALLY ON PAR

In chapter 4, I address time's relationship to minds by considering the nature of time-in-thought as distinct from duration. I argue that time-in-thought is an innate idea by first explaining the philosophical reasons why Descartes *should think* it is innate, and then presenting the compelling textual evidence that Descartes *actually did think* it was innate. I argue that Descartes should think it is innate by first arguing that it must be the idea by which one measures and then by arguing that this idea could be neither fictitious nor adventitious. I then note a number of textual passages where Descartes indicates his assumption of an innate idea of time. In arguing that time-in-thought is an innate idea, I establish it to be an attribute that is on an ontological par with duration. Thus, I indicate how Descartes' temporal dualism diverges from that of Suarez. Unlike Suarez's extrinsic time, Descartes' time-in-thought is not simply a constructed idea. Rather, it is an innate idea and thereby an idea of a true and immutable nature. As such, time-in-thought is only "weakly dependent" on minds since it is found in minds but is not created by them. Thus, by establishing that Descartes' two temporal attributes are ontologically on par, I establish the truly unique character of Descartes temporal dualism.

CHAPTER FIVE: TEMPORAL DUALISM AS AN ELEGANT SOLUTION

In chapter 5, I apply my conclusions from the former four chapters to some of the enduring puzzles that have plagued Cartesian scholarship on time; namely, problems with reconciling Descartes' "parts of time" language with his rejection of atoms and voids, and his commitment to the immutability of God. I argue that reading Descartes as a temporal dualist allows me to resolve a number of these enduring puzzles. In general, I argue that the continuity of duration is ontologically prior to its having successive parts, since it is continuous in its creation, but is divided into parts only as a joint consequence of its being conceived *via* time-in-thought and of God's capacity to accomplish whatever can be clearly and distinctly conceived. Insofar as temporal dualism provides an elegant solution for a number of enduring problems in the scholarship, I argue that this fact itself provides a final and substantial piece of support for my interpretation as a whole.

NOTES

1. Most such discussions focus on trying to determine if Descartes' account of time is continuous or discontinuous. Some significant discussions on this debate include: Jean-Marie Beyssade, *La Philosophie première de Descartes* (Paris: Glammarion, 1979); Martial Guéroult, *Descartes' Philosophy Interpreted According to the Order of Reasons: The Soul and God vol. 1*, ed. and trans. Roger Ariew (Minneapolis: University of Minnesota Press, 1984): Chapter 6; Daniel Garber, *Descartes Embodied: Reading Cartesian Philosophy through Cartesian Science* (Cambridge: Cambridge University Press, 2001), and Daniel Garber, *Descartes' Metaphysical Physics* (Chicago: The University of Chicago Press, 1992); Jorge Secada, "Descartes on Time and Causality," *The Philosophical Review* 99 (1990): 45–72; Richard Arthur, "Continuous Creation, Continuous Time: A Refutation of the Alleged Discontinuity of Cartesian Time," *Journal of the History of Philosophy* 25 (1988): 349–375; Bennett, Jonathan. *Learning from Six Philosophers vol. 1* (Oxford: Clarendon Press, 2001); Clarence Bonnen and Daniel E. Flage, "Descartes: The Matter of Time," *International Studies in Philosophy* 32 (2000); Ken Levy, "Is Descartes a Temporal Atomist?" *British Journal for the History of Philosophy* 13 (2005): 627–74.
2. See *Principles* II.23 and 25: AT VIIIA 52–54; CSM I.232–233. Descartes elaborates in *Principles II.25*, "by 'one body' or 'one piece of matter' I mean whatever is transferred at a given time."
3. *Principle II.39*: "It is true that no motion takes place in a single instant of time." AT VIIIA 64; CSM 242.
4. Descartes argues that by considering "the nature of time" it becomes apparent that each subsequent moment of a thing's duration requires the same amount of causal efficacy to keep that thing in existence as would be required if the thing were not yet in existence. This is because time is such that "its parts are not mutually dependent, and never coexist" (AT 13; CSM I.200). As such, one's existence at a subsequent moment cannot be accounted for by appealing to one's existence in a previous moment since Descartes denies the possibility of a cause acting over a temporal distance. See Geoffrey Gorham,"Cartesian Causation: Continuous, Instantaneous, Overdetermined," *Journal of the History of Philosophy* 42 (2004): 389–423 for a good discussion of this topic.
5. The Kantian-link is explained in the last chapter.

ONE

Pre-Cartesian Approaches to Time

An unfortunate hurdle in understanding Descartes' account of time is that his own discussions are scarce and seemingly under-developed. It seems, however, that such brevity might be excused if Descartes simply presumed that his readers had a working familiarity with the major historical approaches to the nature of time. If Descartes presumed that his contemporaries were well-familiar with the general landscape on how to understand time, then it seems he could have simply left markers to indicate how his views align with the broader intellectual landscape. Rather than "reinvent the wheel," if the historical discussions on time all followed a similar mode of conceiving, then Descartes might be excused if he likewise assumed this general structure and simply located his account within it. Indeed, if Descartes thought he was making a radical break from the traditional ways of conceiving of time *then* one might have expected him to say more about his account. The very brevity of his discussions thus suggests grounds for thinking that Descartes did not think this view was particularly novel or inaccessible to his readers.

If Descartes' account of time does presuppose the intellectual structure of his times, then this both explains and (somewhat) excuses his brevity on the topic. Moreover, it also suggests a method for interpreting his account. If Descartes assumed that his readers were familiar with the standard approaches to time, and thus simply located his own account within this basic structure, then it seems that a modern interpreter must be equally familiar with these structures. Accordingly, before looking at Descartes' own claims about time, my first chapter will outline the historical framework that I believe his discussions presuppose.

I will attempt to sketch this framework by examining the major issues and approaches to time advocated by Descartes' predecessors and contemporaries. It is neither feasible nor even necessary to offer a detailed account of these different approaches since it is unlikely that Descartes would have expected his readers to have exhaustive knowledge of each of these views. Insofar as I do not think an exhaustive analysis of these former accounts is needed, I will just briefly present these accounts, and will rely heavily on secondary accounts when doing so. As my goal is simply to offer a quick grounding on the intellectual backdrop in which Descartes was imbedded, I will (like Descartes) not attempt to re-invent the wheel when analyzing the accounts. As will quickly become evident, even a cursory examination of this history indicates how the various accounts tended to offer nuanced versions of broadly similar types. Given this very common framework, it would have been reasonable for Descartes to have assumed that his readers would have a passing familiarity with the general types. I will attempt to provide a contemporary reader with this same passing familiarity.

Among Descartes' major predecessors and contemporaries, I have isolated four particularly important philosophical traditions concerning the nature of time, which I am calling: the Absolutist Approach, the Reductive Approach, the Mind-Dependent Approach, and the Dualist Approach. On the basis of this examination, I isolate three important questions concerning the nature of time which all of these traditions address in one way or another. I contend that the ways in which the previous authors responded to these questions determine fundamental features of their accounts. Given the historical importance of these questions, I will then ask these same questions (in the chapters to follow) of Descartes' account. The three questions I will isolate are:

1. Is time an independent or dependent entity?
2. What is the relation between time and motion?
3. What is the relationship between time and minds?

As a result of this analysis, I will be able to show how Descartes' brief remarks on the nature of time—when studied in an historically sensitive way—can finally be understood as making important statements, which place him in agreement with certain aspects of these major approaches and in opposition to others.

FIRST TRADITION: ABSOLUTIST APPROACHES TO TIME

According to an Absolutist Approach, time is an independent entity that cannot be ontologically reduced to anything more basic. This approach can be traced back to the ancient view that Julia Annas describes as the commonsense, Platonist view that time exists as a "mysterious cosmic entity or container."[1] Whether Plato himself should be read as having offered an absolutist account of time is not obvious, though this suggestion has been offered by some.[2] Even if one does not agree that Plato himself offered an absolutist account of time, it is evident that such a view was around in the ancient world because Aristotle offers his own view on time in opposition to such a view. Looking at Aristotle's implied target is thus useful for sketching an initial picture of an absolutist approach to time.

Aristotle's approach to time, explains Annas, was motivated by his general discomfort with positing the existence of abstract objects that exist apart from concrete objects. The idea of a mysterious entity existing independently of concrete instances of timings "is already obnoxious for Aristotle, since it would mean that our statements about time were made true by an entity mysterious to us, over and beyond what we can describe in statements whose truth-conditions are given by activities like timings."[3] As Aristotle opposed the absolutist view by himself linking time to concrete objects, his target appears to separate an account of time from an account of concrete objects or timings. Thus, the "obnoxious" subject of Aristotle's critique appears to be anyone who would argue that time possesses an independent existence of its own.

It seems that Aristotle's Platonist target affirms an account of time that follows some of the common idioms used to describe it. Common idioms that refer to events occurring *within* time would seem to imply that time exists as a type of cosmic container within which events occur. Accordingly, this absolutist (or container) view is a "commonsense view"—one that posits time to be something that exists independently from, and evidently prior to, those things that it contains. In order to exist as the container within which events *can* endure, it cannot be the case that the container only exists in virtue of the things that it contains.

There is ample evidence that absolutist accounts of time survived Aristotle's reductivist attack upon them. One such account appears in the thought of Strato of Lampasces, who was the second successor of Aristotle at the Lyceum (from 233 to c. 269 b.c.e).[4] Strato describes time as being some quantity that contains motions. In *The Physical World of Late Antiquity*, Sambursky quotes Strato's account of time as given by Simplicius in his commentary on the *Physics*.

> Action and motion are quicker and slower, but the quantity in which action takes place is neither quicker nor slower but is more or less, and this is time. Day and night, a month and a year are not time, but they are light and darkness and the revolutions of the moon and the sun. Time, however, is a quantity in which all of these are contained.[5]

In defining time as being different from, and the container for cosmic motions, Strato indicates his acceptance of an absolute account of time. Strato seemed to believe that time needed be distinct from the various motions it contains in order to possess the uniform, constant flux needed to measure the various motions contained within it.[6]

Though there are others who seem to offer absolutist accounts of time after Strato,[7] the most interesting historical figure to suggest such a view (interesting in terms of his potential for influence on Descartes) is Pierre Gassendi (1592–1655). Gassendi not only frequently corresponded with Descartes, but he also challenged Descartes on the subject of time. Gassendi's own account of time is given in his *Syntagma Philosophicum*.

> Our opinion is that even if there were no bodies there would be both a constant Place and a flowing Time; whence it is contended that Place and Time do not depend upon bodies and are not corporeal accidents. But neither are they incorporeal accidents…they are, rather, certain incorporeal entities which differ in genus from what are usually termed substances and accidents…even if this or that substance or accident would perish, nonetheless Place would continue to be and Time to flow. Whence it is that place and time are to be regarded as true things and as entirely real...Even though...substance and accident did not exist, nevertheless [Place and Time] would exist: neither do they [i.e., Place and Time] depend upon the Intellect as Chimeras do.[8]

Given that Gassendi proposes time to be an "entirely real" entity, this text clearly indicates that even among Descartes' most immediate contemporaries there were persons positing that time might exist as a something free from dependence on the things that endure through it. Not only did Gassendi deny that time was dependent on some substances or attributes, he was willing to grant that time exists before and after such things. He claims, "time always elapses at the same rate whether anything endures in it or not."[9] Accordingly, Gassendi denies that the class of "being" is limited to substances and attributes as "common opinion" claims.[10] Rather, he claims that place and time are additional sorts of beings that exist independently from substances and attributes, which enable them to measure those things that endure within them.[11]

From Gassendi alone one gets both a picture of an absolute approach to time, as well as strong grounds for assuming that this picture was one with which Descartes was familiar. According to Gassendi's account, time and place "are to be regarded as true things" that are not dependent on minds and that would exist irrespective of substance or accidents. As such, Gassendi grants time the sort of independence toward which Aristotle's much earlier critique was directed. Thus, it is evident that among at least one of Descartes' closest contemporaries, there was a well-developed account of time that followed the absolutist approach.

We can now summarize different common characteristics of the absolutist traditions found in Plato, Strato, and Gassendi using the three questions from above.

1. *Is time an independent or dependent entity?* According to the absolutist, time is an independent entity.

2. *What is the relation between time and motion?* According to the absolutist, time is ontologically prior to motion and a "container" within which motions exist and are measured.
3. *What is the relation between time and minds?* According to the absolutist, time exists independent of minds. Minds exist within time and time could exist even if there were no minds.

TRADITION TWO: REDUCTIVE APPROACHES TO TIME

Although Gassendi's position indicates that absolutist accounts of time endured well beyond Plato and Strato, there is also ample evidence which indicates that reductive approaches to time were the dominant position among the Pre-Cartesian philosophical traditions. According to the reductive approach, time is not an ontologically basic entity, but is instead somehow dependent upon something more fundamental. This reductive approach was favored not only by Aristotle, but also by many of the scholastics that followed him. In briefly outlining Aristotle's reductive account (as it appears in his own works and in some of his followers), one finds both the motivations for the approach, as well as its two most common forms: celestial reductionism and relationalism.

Aristotle's extended discussion of time occurs in the *Physics IV*, where he defines time as being "the number of motion in respect of 'before' and 'after.'"[12] As this definition indicates, Aristotle proposed an intimate link between time and motion. Aristotle connected time to motion and therefore to the activities of the ordinary constituents of the physical world by identifying time with the 'before' and 'after' that a soul is able to apprehend by way of changes or motions. The difference that one is able to note by way of these motions allows one to enumerate a diverse collection of "nows." Aristotle contends that the enumeration of these nows, that is, of the various states of a thing in motion, is just what time is. By identifying time with the enumeration of progressive states of physical objects, Aristotle's account of time clearly contrasts with the absolutist picture. Aristotle suggests that time is nothing other than the motions of the physical world insofar as they give rise to an enumeration by way of rational souls noticing the changing states. Thus, Aristotle contends that one need only study the workings of concrete particulars in order to understand the nature of time. Accordingly, he denies the necessity of positing the existence of some additional, independently existing temporal container.

Though many who followed Aristotle approved of his general reductive approach to time, they disagreed about what time is, or where it is found, within the Aristotelian picture. One of the reductive strategies that received significant support is the view that time reduces to specific celestial motions, that is, "celestial reductionism." In his article entitled

"Celestial Reductionism," Piero Ariotti argues that this view was the "prevalent conception of time" following Aristotle.[13] Ariotti defines celestial reductionism as "the view that conceives time as owing its existence and particular nature to some well specified actual motions."[14] Ariotti argues that this was in fact the view of Aristotle himself, though this point is controversial.[15] Whether or not Aristotle himself maintained a version of celestial reductionism, some of his predecessors appear to have done so.

In the *Summa,* Aquinas appears to offer a version of celestial reductionism when he identifies time with an attribute of the first motion.

> The true reason why time is one, is to be found in the oneness of the first movement by which, since it is most simple, all other movements are measured, as it is said in *Metaphysics X*. Therefore, time is referred to that movement, not only as a measure is to the thing measured, but also as accident is to subject; and thus receives its unity from it. Whereas to other movements it is compared only as the measure is to the thing measured. Hence, it is not multiplied by their multitude, because many things can be measured by one separate measure.[16]

In this text, Aquinas claims that the numbering of the first motion provides a standard according to which all other motions can be compared and counted. In other words, Aquinas attempts to grant the Aristotelian suggestion that time reduces to the number of motions without thereby conceding a multiplicity of times. This worry about "multiple" times is the natural consequent of the Aristotelian move of locating time within the motions of particular existing substances. Insofar as there are numerous movers, a reductivist following Aristotle must offer some explanation for how a multiplicity of concrete movers could still account for the existence of a singular time according to which all motions might be compared.

Aquinas noted the multiplicity of reductive times when he asserts that "number, as it exists in the thing numbered, is not the same for all; but is diverse for diverse things."[17] Accordingly, Aquinas explains that if time were merely the number of diverse motions, then there would a multiplicity of different times corresponding to the number of each individual motion. Rather than admitting such a multiplicity, Aquinas suggests that there is a unified time provided by the first motion (as a substance to its attributes), because the enumeration of this most simple motion provides a standard measure according to which the enumerations of all other motions can be compared.

In claiming that time is referred to the first movement, Aquinas not only offers a type of celestial reductionism, but he also explains how this reduction makes time immediately dependent on the first motion. In his *Commentary on the Physics* at 4.17 [574] Aquinas explains, "Whatever is mutable in existence is such because of that first motion, which is the motion of the first mobile object . . . time is consequent upon only the one first motion by which all other motions are caused and measured."[18] Given the claim that the first motion is both the cause and the measure of other motions it follows that the first motion is not merely the organizing principle according to which the times of various motions are *conceived* in a unified way. Rather, there would *be* no other motions without the first motion, and thus nothing temporal without this first motion. Time is a direct consequence of this first motion, and thus dependent on this particular motion in the same way that all other motions are dependent on the first motion as their cause. Time has an even more intimate link to the first motion, however, because time is not only caused by this motion, but it is also identical with the measure of this first motion.

Aquinas's example shows how identifying time with the measure of specific cosmic motions enabled celestial reductionists to not only provide singular, reductive accounts of time, but also to explain how time is both continuous and regular. Time is continuous and regular because the cosmic motions are continuous and regular.

> Since motion in respect to place is motion from something to something in respect to magnitude, and since every magnitude is continuous, then it is necessary that motion is consequent upon magnitude in continuity. That is since magnitude is continuous, motion is continuous, and consequently time is also continuous.[19]

Celestial reductionism thus provides an elegant solution to a number of problems which all temporal reductivists must face: It explains how time can be singular despite the fact that there are multiple motions, and it explains how time can be both continuous and regular. Despite these attractive features of celestial reductionism, the view had its critics. In the *Confessions*, for example, Augustine explicitly rejects celestial reductionism. Augustine claims, "I have heard from a certain learned man that the movements of the sun, moon, and stars constitute time, but I did not agree with him."[20] Augustine proceeds to argue against celestial reductionism on the theological grounds that it would conflict with Jos.10:12, in which it is recorded that God stopped the motion of the sun and moon and yet activity (and thereby time) continued.[21]

Though Augustine dismisses celestial reductionism as being both naïve and inconsistent with scripture, both Augustine and Aquinas's discussions of the view show the endurance and the appeal of the approach. For those inclined to follow Aristotle's lead in conceiving of time as being something based in the concrete world, and not as something existing as a mysterious, cosmic container, celestial reductionism offered an appealing option. Not only did this account locate time within the activity of concrete beings (namely, in celestial bodies), but since the cosmic motions were continuous and uniform, time was thereby also given a continuous and uniform character. The time of celestial reductionists earned the characteristic benefits associated with the celestial motions while also seeming to cohere with the commonsense notion that time is determined by the sun's rising and setting.

Though celestial reductionism may have been one popular reductive strategy, there were others who accepted Aristotle's reductive approach without arguing that time was identical to specific motions. This group, who I am calling the relationalists, denied that time was identical with motions *per se*. Instead, relationalists claimed that time is a relation that depended on motion as its *relata*. In the *Physical World of Late Antiquity,* Sambursky identifies Aristotle himself as a relationalist.

> [Aristotle] clearly recognizes that time and motion are interrelated. In the same way that space exists only in so far as there are bodies which occupy a certain place, time exists only in so far as there are bodies which, at different 'nows,' are in different places or in different states.[22]

Though this relationalist approach accords with the reductivist desire to reduce discussions of time to discussions of events occurring within the physical world, they posited a more complex entity (i.e., a relation) as the thing to which time is reduced.

In his analysis of the reductive strategies of thirteenth-century thinkers, Rory Fox argues that a relationalist reduction was the only type admitted by Aristotle, because Aristotle had ruled out reducing time to motion in "crude simplistic ways." Fox argues that Aristotle had ruled out such reductions because time and motion possess divergent properties; namely, time is everywhere, whereas motion is in specific things, and motion can change velocities whereas time cannot.[23] A relationalist approach distinguishes itself from the "cruder" celestial reductionist account because the relationalist account only identifies time with one particular feature of motion and not with motion *per se*. The relationalist reduces time to that particular feature of motions whereby they indicate difference when they are compared to other motions. Likewise, the relationalist account distinguishes itself by identifying time with this particular feature as it is given in all motions, and not merely when it is noted in specific, cosmic motions. This subtler account seems to have been given its most careful presentation in Epicurus.

From what is known of Epicurus's account, it is evident that his account was reductive since it reduces discussions of time to discussions of relations. However, because Epicurus suggests that relations are not ontologically basic, his account does not reduce time to a basic ontological category. In *Time in Late Antiquity,* Ariotti claims that Empiricus and Laertius (the two primary sources of Epicurus's account of time) "agree that for Epicurus time is neither an independent existent, nor a result of motion or chance, nor an essential characteristic of existents."[24] Epicurus's view, according to Ariotti, is that time is a "second order property or attribute of world existents."[25] By this he means that time is neither an independently existing constituent of reality, nor even a basic attribute of such a body. Rather, time is an attribute of processes. Laertius writes that

> ...we attach the attribute of time to days and nights and their parts, and likewise to feelings of pleasure and pain and to neutral states, to states of movement and states of rest, conceiving a peculiar accident of these to be this very characteristic which we express by the word "time."[26]

In describing time to be an "accident" or "attribute" of process, the Epicurean account that Laertius explains is that time is neither a true existent (i.e., a body), nor even one of the fundamental properties of a body (e.g., shape, color, magnitude, weight), but rather a non-permanent quality of bodies that is not essential to their natures. According to this framework, only bodies are true existents whereas the properties/attributes of bodies are also features of reality, but features of lower ontological standing. The fundamental properties that are permanently conjoined with bodies possess weaker ontological standing than bodies because they depend on bodies for their existence. Though time is like these properties in being likewise dependent on bodies, time is assigned an even lower ontological standing than these properties because it is not (like they are) a permanent property of bodies. Time is not a permanent property of bodies because it is possible to conceive of bodies without conceiving of time.

In Ariotti's attempt to elaborate on what Epicurus might have meant in designating time to be a non-permanent property of bodies, he contends that Epicurus may be suggesting that "time is a relation which we see or use in the comparison of motions."[27] According to this account, time is not found in the motions of bodies *per se*, but rather, time is found in the comparisons or relations that obtain between various motions. An important point to note is that if Ariotti is correct in suggesting that Epicurus conceived time to be a relation that "we see or *use*" then this interpretation points to another question to ask of Epicurus' account: What role do minds play in this account? In *Time and Mind,* J.J.A. Mooji wonders if according to Epicurus, time might be dependent on mind[28] and Ariotti's analysis of Epicurus does suggest some grounds for supposing an affirmative answer to this query. If time is a relation that is seen or used, then it seems that this account might posit time to be dependent on a mind to do this see*ing* or us*ing*. As the next section will show that there were many who did suppose that time was somehow dependent on minds, this additional possibility might be important for fully understanding not only Epicurus's account, but the relation between these standard views in the reductivist tradition and the mind-dependent tradition (which is the subject of the next section.) Because the texts on Epicurus's view are so few, it is sufficient to note that this possibility arises within his account, and then to postpone an analysis of the role of minds in various accounts of time until the next section.

Even the very brief sketch of Epicurus's view that I have offered is sufficient for recognizing how the relationalist reductivist account of time offers a third perspective on the reductive v. absolute account of time debates prior to Descartes. Though relationalists certainly side with the reductivist's position that time is not an independently existing constitute of reality, they offer a more complex reduction by positing that time reduces to a relation, and thus to a second-order property of bodies. As such, this view posits that time is neither substance, nor a fundamental attribute of substance. That is to say, time is neither substance, nor the motion of a particular substance, but rather, a relation that is found between various motions. To sum up, these standard kinds of reductivists answer the three central questions concerning the nature of time as follows:

1. *Is time an independent or dependent entity?* According to both the celestial reductionist and relationalist, time is a dependent entity. The celestial reductionist argues that time is identical with a specific cosmic motion (thus accounting for the unity of time) whereas the relationalist argues that time is identical with a relation between motions.
2. *What is the relation between time and motion?* According to both the celestial reductionist and the relationist, time is dependent on motion for its existence. They disagree about how time depends upon motion, however. According to the celestial reductionist, time is directly dependent on one particular motion. According to the relationalist, time is dependent upon motion as the *relata* that gives rise to time (the relation between motions).
3. *What is the relation between time and minds?* Here we get the biggest difference between the two branches of reductionism. According to celestial reductionism, time is not dependent upon minds. According to at least some relationalist accounts (perhaps, for example, that of Epicurus) time might be dependent on minds for its existence if time is interpreted as being identical not to the relation between motions themselves, but rather to the *conceiving* of this relation.

TRADITION THREE: MIND-DEPENDENT APPROACHES TO TIME

Though Epicurus's account of time may suggest some level of mind-dependence, there were many others who not only included some kind of dependence on minds, but who made this mind-dependence central to their approach. In this section, I will explore influential examples of those whom I identify as proponents of the mind-dependent approach to time. Insofar as these approaches describe time as being mind-*dependent,* they might appropriately be classed among the broader heading of reductivists. As, however, they offer a different sort of reduction (i.e., one that reduces motion to features of minds rather than to concrete objects or their activities), they will be considered as offering a distinct approach to time.

The notion that the nature of time was intimately tied up with the nature of minds was a recurring theme among many of Descartes' predecessors. Even Aristotle had identified an important role for minds in his account of time. By defining time to be the number or measure of motion, Aristotle suggested that time was dependent on minds since "if there cannot be some one to count there cannot be anything that can be counted either."[29] Accordingly, it is not surprising that many of Aristotle's followers shared the belief that time is somehow dependent on minds.

In *The Physical World of Late Antiquity,* Sambursky claims that considerations of time's relationship to change led many to suppose that time might be dependent on minds. Sambursky explains how the "problem of time and consciousness" naturally arose from such considerations since time is "invariably connected with consciousness, with the apprehension of an eternal and continual flux . . . the very flow of time is bound up for us with the experience of the flow in our consciousness."[30] Given this link, it is not surprising that one can find various thinkers giving minds central roles in their accounts of time. Plotinus, for example, claims that "time is the life of the soul as it moves from one state of life to another."[31] Likewise, In book 11 of the *Confessions,* Augustine suggests that time is a distention of the mind itself.[32] Given that Augustine's various cogito-like arguments are widely regarded as having greatly influenced Descartes' thought,[33] Augustine's explicitly mind-dependent account of time seems particularly important to study for the possible ways that Augustine's views on time may have influenced those of Descartes.

Augustine develops his mind-dependent account of time as a solution to certain of the paradoxes of time that had been posed by predecessors. He not only addresses the ancient worry of Aristotle that time might not exist at all,[34] but he also addresses a theological question that had received some attention; namely, he addresses the question of what God was doing "before" he chose to create. To the question of what God was doing before he chose to create, Augustine denies the question's presupposition that there was time before creation. Indeed, Augustine argues that "there can be no time without creation"[35] since time just is the mode of being for creation, whereas eternity is the mode of being for the creator. To the Aristotelian worry that time might not exist at all since "one part of it has been and is not, while the other is going to be and is not yet,"[36] Augustine offers a more extensive response. Augustine focuses on the question of time's seeming unreality because this question was one that he himself found perplexing.

Augustine worries that time appears to have a very tenuous existence given that time consists merely of an instantaneous present poised between the apparent non-being of the past and future. As Augustine phrases the worry, "the past no longer is and the future as yet does not be."[37] Such an account of the past and future suggests significant worries about the possible being of time given that Augustine also claims that "[The present] flies with such speed from the future into the past that it cannot be extended by even a trifling amount. For if it is extended, it is divided into past and future."[38] If neither the past nor the future exist, and the present is fleetingly small, then Augustine is troubled by how one can make sense of the many accounts of time that describe it as not only existing, but existing according to a longer or shorter expanse. It is primarily to address this last, particularly sticky puzzle that Augustine offers his mind-dependent account of time.

Though it was earlier noted that Augustine explicitly rejected celestial reductionism, he did see an important relationship between time and the motion of the sun; namely, Augustine suggests that time is the distention within which this motion takes place. That is to say, time is the extended period within which these motions occur since Augustine affirms that "a body is never moved except in time."[39] In suggesting that the distention of time is required for the possibility of any motion, Augustine needed to account for how time might be extended in such a way that it might encompass the various stages of a body's motion. Augustine recognized a need to offer some account of how time can possibly be extended over some length despite its seeming to be the progression between "that which does not yet exist, by that which lacks space, and into that which no longer exists."[40]

As Augustine is committed to the claim that existence is only a feature of the present, to explain how time can be extended he must explain how present existence can extend into the non-presentness of the past and future. Augustine achieves this feat by identifying the past, present, and future with features of minds. He identifies the past with one's memories, the future with one's expectations, and the present with one's immediate attention. As one can focus immediate attention on things stored in one's memory or on things that one anticipates, the mind is able to extend present existence to things that are past and things that are in the future. Thus, Augustine claims, "it might properly be said that there are three times, the present of things past, the present of things present, and the present of things future. These three are in the soul."[41]

Augustine argued for the mind-dependence of time in order to extend the present in this way because such an extension would allow a body's motion or rest to be measured. As motion and rest both endure from the past through the present, and into the future, any measure of the distention through which these occur must likewise have some kind of being that endures through the past, present, and future. As existence is a feature of the present, and because a mind is able to bring the things that it recalls and the things that it anticipates into its immediate attention, the mind is uniquely capable of offering a type of present existence to the past and future. As a consequence, Augustine claims, "Time is nothing more than distention . . . and the marvel is, if it be not of the mind itself."[42] Time is the distention of the mind itself because it is only within minds that the past and present have any being. As there could be no measure composed of a past and present without being, the only way that there is a distention through which an enduring body might be measured is if the distention through which bodies are measured is found in minds. As time is the distention through which bodies are measured, time is found in, and wholly dependent on, minds.

Though Augustine may have offered the most explicitly psychological account of time among Descartes' predecessors, the idea that time was dependent on minds was a recurring theme. Whereas few went as far as Augustine's claims that time is identical with a distention of mind, there were many who took a more Aristotelian stance that time requires minds as the means of uncovering the more fundamental temporal features in the world. Among those who gave time some level of mind-dependence, there was one additional group that has not yet been considered. There were some who suggested that time might be mind-dependent because it exists as a type of an *idea* in minds. This last group tended to argue that time is not merely an idea, but that it also exists in some form in the physical world. This last group, who I will identify as temporal dualists, will be considered at greater length in the following section.

To sum up in this section, mind-dependent theorists answer the three important questions in the following way:

1. *Is time an independent or dependent entity?* According to the mind-dependent theorist, time is a dependent entity which requires minds in order to exist.
2. *What is the relation between time and motion?* According to the mind-dependent theorist, time is distinct from corporeal motion, but depends upon the conceiving of change that occurs within minds.
3. *What is the relation between time and minds?* According to the mind-dependent theorist, time is dependent upon minds for its existence.

TRADITION FOUR: DUALIST APPROACHES TO TIME

The last approach to time which received significant support among Descartes' predecessors and contemporaries is what I am calling the dualist approach.[43] According to the dualist approach, there is more than one kind of time, with one kind generally serving as the measure of the other. I am examining this approach last because the dualist approach presupposes a familiarity with absolutist, reductive, and mind-dependent approaches to time. Thus, a temporal dualist might claim that time both is and is not mind-dependent—depending on which kind of time is under consideration.

To a certain extent, one might argue that many (perhaps even most) of Descartes' predecessors were temporal dualists. One might think that temporal dualism was the norm insofar as it was common to affirm the separate existence of time and eternity—with time characterizing the mode of existence of creatures and eternity characterizing the mode of existence of the creator. Though it's true that the time/eternity divide was pervasive, the particular dualism that I intend is not a distinction between time and eternity. Rather, I am considering a distinction that is only found in discussions about time as conceived apart from eternity. Thus, the fact that various theorists maintained a distinction between time and eternity is not by itself sufficient for their being included among the class of temporal dualists. It also follows that persons classified as temporal dualists may offer more than two temporal accounts if one includes discussions of eternity in the analysis.

In *Time, Creation and the Continuum*, Richard Sorabji offers the Neoplatonic philosopher Iamblichus (C.E. 250–325) as an early example of a temporal dualist. Sorabji explains that Iamblichus "treats eternity as distinct from his two kinds of time, since he draws a threefold distinction among them."[44] The two kinds of time that Iamblichus identifies are distinguished as "higher" and "lower." As their names suggest, Iamblichus does not give these two types equal ontological status. Rather, Sorabji explains how Iamblichus offers two types of time that appear to map on to a platonic form of time and the time that participates in the form.[45] Iamblichus's "lower time" is composed of an ever-changing now. It is always in a state of coming to be rather than in a state of static existence, whereas his "higher time" is the "image of eternity." As the image of eternity, this higher time endures as a static whole akin to a platonic form.[46]

By positing a higher/static time as participated in by the lower/flowing time, Iamblichus gained a solution to the enduring puzzles about the reality of time and provided an explanation for how time could serve as a measure. Lower time is the time of things that endure successively. The time of these flowing, changing entities is a time that would suffer the same tendencies towards non-being that have already been considered; namely, in virtue of the flowing, the lesser time is always merely in a state of becoming rather than in a state of being. Iamblichus's solution to this problem was to argue that this lower time has its being in virtue of its participating in the static being of the now of the higher time.

> ...there is a difference between the now in which things participate in nature, and which is inseparable from things which come to be, and the

> [now] which is separate and by itself. One of these stands still in the same form consistently; the other is seen in continuous motion.[47]

By arguing that there are two "nows" in time—the now of lower time and the now of higher time—Iamblichus can ground the being of time in the higher now, while conceding the flowing nature of lower time. In positing these two kinds of time, one can see how Iamblichus's temporal dualism has links to both absolutist and reductive accounts of time. Whereas "higher time" has characteristics in common within an independent, absolutist account, "lower time" accords with the dependent, reductivist accounts.

A significant advantage to Iamblichus's dualist picture is that it allows him to account for how time can serve as a measure. Sorabji notes that Iamblichus's dualist distinction in time was upheld by Neoplatonists following Iamblichus because his account explained how time could provide a measure for those entities enduring through the flowing stages of lower time. Since time could not serve as a measure of such flowing if it simply "kept pace" with the flowing, there needed to be an external, static time to enable this measurement. As Sorabji explains, "if moving time is to be measured, there will have to be something stationary to measure it."[48] Iamblicus's temporal dualism possesses this stationary character in virtue of the higher time which grounds it.

Though Iamblichus's account of time indicates him to be one sort of temporal dualist, his account does not seem to commit him to the view that one of these types is an idea dependent on minds (though it is not perfectly clear what type of being his higher time might have possessed).[49] As I earlier indicated, however, there were certain temporal dualists who did maintain that the "measuring time" was a mind-dependent idea. In the *Disputationes Metaphysicae*, for example, Suarez presented a dualist theory of time, but he divided these two types according to their kinds of dependence. Whereas one kind depended on the existence of enduring substances, the other was a mind-dependent idea. Suarez's account is a particularly important account in the present context because the *Disputationes* were so widely used during the seventeenth century.[50] Likewise, the fact that Descartes makes explicit reference to the *Disputationes* makes it very likely that this text was part of the background knowledge that Descartes employed and presumed.[51]

Suarez's account of time develops between the fortieth and fiftieth disputations.[52] Within these disputations, Suarez indicates that there are two types of time which (similar to Iamblichus's account) differ in their ontological standing. Unlike Iamblichus, Suarez reverses the order of priority between the two types. Whereas Iamblichus had argued that the static, complete time was more ontologically basic than the flowing time found in the enduring of created substances, Suarez maintains that the fundamental nature of time is found in these successive endurings.

Though Suarez locates time within this succession, he does not claim that time is some sort of empty timeline in a constant state of becoming. Rather, time is fundamentally tied to being, since time just is the successive enduring of beings in motion.

> [...] there is no distinction in a thing between [its] duration [...] and that existence of which it is the duration [...] For in the same motion, for example in one circulation of heaven, there is the same real duration, because it is the same real existence of such a motion [...].[53]

By defining time as the successive enduring of beings in motion, Suarez has motion play an important role in his definition of time; namely, motion actualizes the potencies of beings to be described numerically. As Suarez claims, "[Time] is not simply the number of a continuum, but the number of the parts of the same continuum, which in the thing is only number in potency, actualized through the operation and numeration of the mind."[54] As this passage indicates, motion is necessary for bringing the temporal count of a thing from existing in potency to actuality. Motion seems to serve this role in a similar way for Suarez as it did for Aristotle. Motion allows a mind to recognize when a thing has transitioned from before to after, and in recognizing this difference the mind is able to apply a number to the enduring of the thing.

Though apprehending motions allows a mind to count a thing's enduring, it is important to remember that the count is not time itself. This counting may require an external cause (i.e., a mind), but the thing counted (i.e., the thing's enduring) is not an external or accidental feature of a thing. Rather, time is something contained within the very nature of a thing. Indeed, since Suarez locates time within a thing's very enduring, he frequently refers to time as "intrinsic duration." Given that Suarez's position locates time within the enduring of substances, his account has much in common with the reductive accounts; however, it diverges in virtue of its unusual, dualistic character.[55]

Given that Suarez maintains that the real temporality of enduring substances is found within the successiveness of their natures, this inherent succession could not offer a means of comparing various durations.

> [...] in all continuous motion there is intrinsic and real successive duration, and consequently, if the name of time signifies this kind of duration alone, it follows that there is not one and the same time for all motion, but [that time is] multiplied according to the number and multitude of motions, and is various according to the diversity of motions.[56]

As each substance possesses its own intrinsic duration, such succession only tells us about the number of different stages through which a substance has endured. In counting these stages, one can determine a multiplicity of durations, but this would not provide a means of comparing various durations. In "Seventeenth-Century Scholastic Accounts of Time," Stephen Daniel explains, "the duration of a particular movement is its own time, measurable by itself alone, and completely encompassing the entire extent of its movement."[57] Thus, in order to explain how various durations can be compared, Suarez admits an additional, "extrinsic time" as distinct from this "intrinsic" durational.

The dualistic character of Suarez's account of time thus arises from his need to identify an objective means of measuring the various intrinsic durations. To gain this objective measure, Suarez posits "imaginary time" or "extrinsic time." Imaginary/extrinsic time possesses the successive character of intrinsic durations but lacks their inherence (i.e., it is not found within the duration of any specific substance).[58] Suarez conceived imaginary time as offering an external, objective standard within which real, intrinsic durations coexist in their enduring.[59] This coexistent time provides a second means counting the enduring of a thing. There is the count of the thing's real, intrinsic duration (determined by its own motions), and there is the count of this external measure that can be applied to a thing to get a (possibly) different count. Suarez makes it clear that this count is distinct from the count found in intrinsic duration.

> The duration of a movement can be considered by comparison with, and as coexistent with, an imaginary succession which we apprehend as infinite; and in this sense, the duration which is in a swift movement through an equal space is smaller, and in a slow movement it is longer because it coexists and in a way fills up (as I call it) a greater or lesser part of that imaginary time. But it is not the same thing, for the duration of a movement has another extent or succession apart from the movement itself.[60]

Insofar as the imaginary succession is distinct from a thing's intrinsic duration, the imaginary succession does not tell us about the intrinsic nature of the thing itself. To understand the fundamental nature of a thing one would consider its intrinsic duration. To compare the duration of one thing to another, however, one would look to this second, coexisting succession.

Like intrinsic time, Suarez's imaginary/extrinsic time is tied up with motion. Suarez claims, "the imaginary flowing period is conceived as totally necessary and immutable in its flux."[61] In such phrases, one sees that extrinsic time is characterized by movement (e.g., by a "flowing" or "flux"). This is not surprising, since Suarez is committed to movement being the ground that actualizes the numerical character of any succession; however, this requirement is somewhat problematic. As motion characterizes specific bodies and his imaginary time is supposed to be an external succession applicable to all bodies, Suarez's account of imaginary time must be drawn (in part) from some specific motions, but cannot be confined to these motions. It is here that Suarez appeals to the celestial motions (for much the same reason as the celestial reductionists did). Suarez argues that

> We conclude that time is unique in the universe because it has its own meaning of extrinsic measurement, viz., in the movement of the heavens... Understood as an extrinsic measurement and as common to other movements, it is taken partly from nature as it fundamentally exists, or inchoative (as I call it), and partly from a division through reason and an accommodation of it as a means of measurement; and in this way there is only one time, and it is the movement of the heavens. . . From its very usage it is very well enough known, for however often we want to determine the duration of some action or inferior movement, we compare it with the movement of the heavens so that we might understand it.[62]

Extrinsic time relies on the celestial motions as offering a standard duration that can serve as a common measure for all other intrinsic durations. Like all motions, the motions of the celestial bodies possess an intrinsic duration, and like all other motions, their duration is particular to themselves. However, the numerical ordering of these particular motions can serve as an external measure for all other motions once an enumeration of the motions is abstracted as an idea (e.g., "from nature as it fundamentally exists") and then modified (e.g., partly from a division through reason and an accommodation of it as a means of measurement"). The mind must modify the original idea by adding, for example, that the celestial motions are non-repetitive, necessary, and immutable.[63] By imaginary time existing as a rationally modified idea, Suarez's imaginary time avoids certain limits of celestial reductionism; namely, the succession can be rationally stipulated to be perfectly regular and eternal and thereby avoids questions about when it began, or what if it stopped, slowed, and so forth.

An analysis of Suarez's intrinsic and extrinsic accounts of time makes it evident that the difference between intrinsic and extrinsic duration turns on the distinction between the measure and the thing measured. In his article on Suarez's concept of time, Constantino Esposito highlights this relationship.

> It is true that the measurement of a thing is a cognitive operation extrinsic to the thing itself, nevertheless the measurability of a thing in time coincides with the certainly not extrinsic fact that this thing has a *duration of such a nature* that it can be temporally measured.[64]

In other words, for Suarez, the measure of a thing is always extrinsic to a thing since it requires a mind to determine the measure. The thing measured, however, is not extrinsic to a thing because what is measured is part of a thing's very nature. A similar divide is evident if one looks back at the higher/lower time theory of Iamblichus. Iamblichus's distinction between the measure and the thing measured posited that the measure (higher time) possesses its measuring capacity in virtue of its stable and complete existence, whereas the thing measured (lower time) required an external measure in virtue of its being in the fragile state of becoming. Though the distinction between the measure and thing measured was one discussed by many of Descartes' predecessors, the temporal dualists were unique in positing that time was not merely one or the other. Rather, the temporal dualists posited that time was found both in the measure and in the thing measured.

1. *Is time an independent or dependent entity?* Temporal dualists give different answers to this question. Iamblichus offers one independent kind of time—higher time, and one dependent kind—lower (dependent on higher). Suarez offers two dependent types: intrinsic time which is dependent on the existence of an enduring body (as it is identical to this enduring), and extrinsic time which is dependent on minds to formulate the idea of it.
2. *What is the relation between time and motion?* According to Suarez, both types are dependent on motion to some extent. Intrinsic time depends on motion to actualize the potential of a duration to be conceived numerically; whereas extrinsic time depends on the particular motions of the celestial bodies to give rise to the idea that reason then modifies. According to Iamblichus, neither type is dependent on motion.
3. *What is the relation between time and minds?* According to Suarez, one of the two types is independent (intrinsic time) and the other is dependent (extrinsic time). For Iamblichus, it seems that both are independent of minds.

CONCLUSION

Having briefly outlined four of the major approaches to time offered by Descartes' predecessors and contemporaries, several important points have been made. First, it seems that all of the four traditions were prevalent in the intellectual culture surrounding Descartes to a degree sufficient for them to have influenced Descartes' views. Second, it is clear that all of the various traditions formulated their accounts according to a similar framework of ideas, that is, as ways to address a similar set of questions and problems. Third, three central questions suggest themselves as being particularly relevant to understanding temporal accounts of this time period; namely: (1) Is time an independent or dependent entity? (2) What is the relation between time and motion? And, (3) What is the relation between time and minds? As answers to these questions structured the various accounts of Descartes' predecessors, it seems appropriate to analyze Descartes' account *via* a similar structure.

NOTES

1. Julia Annas, "Aristotle, Number and Time," *The Philosophical Quarterly* 25 (1975): 102.

2. I. M. Crombie, *An Examination of Plato's Doctine's Vol. II: Plato on Knowledge of Reality* (New York: Routledge & Kegan Paul, 1963).

3. Annas, "Aristotle, Number and Time," 102.

4. See Piero Ariotti, "Toward Absolute Time: The Understanding and Refutation of the Aristotelian Conception of Time in the Sixteenth and Seventeenth Centuries," *Annals of Science* 30 (1973): 145; Piero Ariotti, "Celestial Reductionism of Time: On the Scholastic Conception of Time from Albert the Great and Thomas Aquinas to the End of the 16th Century," *Studio Internazionale di Filosofia* 5 (1972): 531–532; Richard Sorabji, *Time Creation and the Continuum: Theories in Antiquity and the Early Middle Ages* (Chicago: University of Chicago Press, 2006), 82; Samuel Sambursky, *The Physical World of Late Antiquity* (Princeton: Princeton University Press, 1962), 10–12; Samuel Sambursky, *Physics of the Stoics* (New York: Macmillan Co., 1959).

5. Sambursky, *Physical World*, 11.

6. Sambursky, *Physics of the Stoics*, 100–101.

7. Including perhaps Boethius, Damascius, Galen, Plotinus, and Bernardino Telesio. See discussions in Ariotti, "Toward Absolute Time," Sambursky, "Physical World," and Sorabji, *Time Creation and the Continuum*.

8. Quoted in Ariotti, "Towards Absolute Time," 160.

9. Pierre Gassendi, *The Selected Works of Pierre Gassendi*, ed. and trans. Craig B. Brush (New York: Johnson Reprint Corp, 1972), 395.

10. Gassendi, *Selected Works*, 384.

11. Gassendi, *Selected Works*, 394ff.

12. *Physics IV.11*; 219b1, in Aristotle, *The Complete Works of Aristotle*, ed. Jonathan Barnes (Princeton: Bollingen Series, 1995), 372.

13. See, for example, Ariotti, "Celestial Reductionism," 91–120.

14. Ariotti, "Celestial Reductionism," 93.

15. When considering the nature of time, Aristotle determines that it must be continuous, uniform, eternal, and universal. Accordingly, Aristotle claims, "regular circular motion is above all else the measure, because the number of this is the best known…This is also why time is thought to be the movement of the sphere, viz. because the other movements are measured by this, and time by this movement."(*Physics IV.14* ; 223b18–2) However, given that Aristotle only here contends that time is "thought to be" this movement, while he previously claimed that time is "simply number of continuous movement, not of any particular kind of it," (*Physics IV.14*; 223a35) it is certainly not obvious that Aristotle himself thought that time was a feature of particular celestial motions.

16. *Summa Theologica* 1.10.6, in Thomas Aquinas, *Summa Theologica*, ed. Anton C. Pegis (Indianapolis: Hackett Publishing, 1945): 82; see also *Commentary on the Physics* at 4.17.

17. Aquinas, *Summa* 1.10.6, 83.

18. Quoted in Rory Fox, *Time and Eternity in Mid-Thirteenth-Century Thought* (New York: Oxford University Press, 2006), 139.

19. *Commentary on the Physics* 4.17: Quoted in Fox, *Time and Eternity*, 140.

20. Augustine, *The Confessions of Saint Augustine*, trans. John K. Ryan (New York: Image Books, 1960): 295.

21. In Jos. 10:12 it is recorded that Joshua prayed for God to halt the sun and moon so that he might achieve success in battle. As the *Bible* records that this prayer was granted, and that these celestial bodies stood still for the length of a day, Augustine infers that clearly time is something more than just the movement of celestial bodies. Augustine claims "that battle was waged and brought to an end during its own tract of time, which was sufficient for it" (Augustine, *Confessions*, 30).

22. Sambursky, *Physical World*, 13.

23. Fox, *Time and Eternity*, 135.
24. Ariotti, *Conception of Time*, 529.
25. Ariotti, *Conception of Time*, 530.
26. Diogenes, *Lives of Eminent Philosophers*, trans. R. D. Hicks (Cambridge: Harvard University Press, 1965): 599. Quoted in Ariotti, *Conception of Time*, 529.
27. Ariotti, *Conception of Time*, 530. This sort of language is quite consistent with the role of time-in-thought that I will later identify and explain as found in Descartes.
28. J.J.A. Mooji, *Time and Mind: The History of a Philosophical Problem* (Boston: Brill Academic Pub, 2005), 39. Likewise, Catherine Rau defines Epicurus's account in mental terms. She claims, "[time] does not exist in itself, but is abstracted from events by the mind." Catherine Rau, "Theories of Time in Ancient Philosophy," *The Philosophical Review* 62 (1953): 518.
29. Aristotle, *Physics* IV.14.223a23, 377.
30. Sambursky, *Physical World*, 16.
31. Quoted in Sambursky, *Physical World*, 163. Sorabji notes that Plotinus's definition referred to the soul of the universe and not to an individual soul, however. See Richard Sorabji, *Time Creation and the Continuum: Theories in Antiquity and the Early Middle Ages* (Chicago: University of Chicago Press, 2006), 94.
32. Augustine, *Confessions* 11.26.33, 298.
33. For a good discussion on this point see: Gareth Matthew, *Thought's Ego in Augustine and Descartes* (Ithaca and London: Cornell University Press, 1992).
34. Aristotle, *Physics* IV.10, 218a2, 370.
35. Augustine, *Confessions* 11.30.40, 303. In such descriptions, one can see how Augustine's mind-dependent account accords with the reductive approaches.
36. Aristotle, *Physics* IV.10, 218a2, 370.
37. Augustine, *Confessions* 11.14.17, 288.
38. Augustine, *Confessions* 11.15.20, 289. In supposing that an extension of time admits division into time, Augustine suggests the sort of reasoning that led others to reject the ideas of temporal atomism. For some discussion on the Stoics' rejection of these ideas see: Sambursky, *Physics of the Stoics*, 103ff.
39. Augustine, *Confessions* 11.23.30: 296.
40. Augustine, *Confessions* 11.21.27: 293.
41. Augustine, *Confessions* 11.20.26: 293.
42. Augustine, *Confessions* 11.26.33: 298.
43. This terminology accords with Emmaline Bexley, "Quasi-Absolute Time in Francisco Suarez's Metaphysical Disputation," *Intellectual History Review* 22 (2012): 5–22.
44. Sorabji, *Time Creation and the Continuum*, 44.
45. Sorabji, *Time Creation and the Continuum*, 38. This sort of account might be conceived as roughly analogous to the McTaggart divide of the B-series from the A-series.
46. Sorabji, *Time Creation and the Continuum*, 38.
47. Quoted in Sorabji, *Time Creation and the Continuum*, 40.
48. Sorabji, *Time Creation and the Continuum*, 42.
49. Sorabji points to a discussion by John Dillon, where Dillon argues that this entity was not actually a type of Platonic form. As something lesser, it is difficult to classify.
50. Stephen H. Daniel, "Seventeenth-Century Scholastic Treatments of Time," *Journal of the History of Ideas* 42 (1981): 589. Though Suarez's view will be considered in detail (because of its potential influence on Descartes), similar dualist accounts were offered by others. In *Descartes and the Last Scholastics*, Roger Ariew identifies Eustachius, De Ceriziers, and Marande as others proposing very similar views. See Roger Ariew, *Descartes and the Last Scholastics* (Ithaca: Cornell University Press, 1999), 53ff.
51. *Fourth Set of Replies:* AT VII 235; CSM II.164.
52. My interpretations of Suarez owe much to the work done by Daniel, "Scholastic Treatments of Time," and Emmaline Bexley, "Quasi-Absolute Time in Francisco Suarez's Metaphysical Disputation," *Intellectual History Review* 22 (2012): 5–22.

53. *Disp. Meta.*, 50.9; Quote and trans. from Bexley, "Quasi-Absolute Time," 8.

54. *Disp. Meta.*, 50.12.9; Quoted in Daniel, "Scholastic Treatments of Time," 592.

55. For a discussion on the reductive nature of Suarez's account, see Ariotti, "Towards Absolute Time," 35ff.

56. *Disp. Meta.*, 50.8.6; Quote and trans. from Bexley, "Quasi-Absolute Time," 9.

57. Daniel, "Scholastic Treatments of Time," 594.

58. What Suarez implied by calling this succession "imaginary" is somewhat controversial. Whereas Daniel would claim that this extrinsic time is imaginary in virtue of being an idea that one fabricates, in her more recent article, Emmaline Bexley argues that "imaginary" is merely Suarez's way of indicating that an extrinsic time (i.e., a time not inherent to an enduring substance) is a thing that can only be conceived in minds because succession without a thing enduring successively is fundamentally a privation. This external time is thus a type of thing in the same way a vacuum is a type of thing. Says Bexley, it is a thing that exists in some sense, since it has a nature that can be truly predicated of, and yet it is a thing that can only be found in minds. Bexley's analysis intends to show that Suarez's extrinsic time suggests an important precursor for the later absolutist accounts of Gassendi or Newton. Though her analysis is quite interesting and somewhat compelling, I tend to favor Daniel's view which interprets Suarez's "imaginary" time with the minimal, ontological standing available for a fabricated idea. My grounds for favoring this view are drawn from the later passages I will be presenting. It is, however, worth looking at Bexley's article to see this alternative interpretation.

59. Daniel, "Scholastic Treatments of Time," 595.

60. *Disp. Meta.* 40.9.10; Quoted in Daniel, "Scholastic Treatments of Time," 594.

61. *Disp. Meta. 50.9.15*; Quoted in Daniel, "Scholastic Treatments of Time," 595.

62. *Disp. Meta. 50.10.11*; Quoted in Daniel "Scholastic Treatments of Time," 596.

63. Daniel, "Scholastic Treatments of Time," 597.

64. Constantino Esposito, "The Concept of Time in the Metaphysics of Suarez," in *The Medieval Concept of Time: Studies on the Scholastic Debate and its Reception in Early Modern Philosophy*, ed. Pasquale Porro (Leiden: Brill, 2001), 394.

TWO

Time is Not (and Could Not Be) a Cartesian Substance

In this chapter, I argue that time is not—and indeed *could not be*—a Cartesian substance. That is to say, I argue that it is an historical fact about Descartes that he didn't think time was a substance, while also identifying the deep metaphysical reasons that prevented him from maintaining such a view. In so doing, I clarify Descartes' account of time by showing how the account is necessitated by his metaphysics, and thus why it is not an *ad hoc* addition. I will do this by (i) clarifying the categories of Descartes' ontology, then (ii) examining the primary texts that appear to suggest the interpretation that time could be a substance, before finally (iii) offering four arguments (both historical and philosophical) against the possibility of time's being a Cartesian substance.

The work in this chapter builds on the historical context sketched in the previous chapter. In particular, this chapter relates to the ancient debate over whether or not time could exist as an independent entity. In chapter 1, I identified the absolutists as answering in the affirmative, and the reductivists in the negative. That is to say, the reductive accounts of time maintained that time is dependent on "more fundamental" features of the world (e.g., the motion or relations obtaining among corporeal objects, minds, etc.) whereas the absolutists argued that time is so radically *in*dependent that it requires nothing else—not even the existence of particular objects existing in time—for its existence.

The historical debate between absolutist and reductive accounts of time manifests itself within the Cartesian context as a debate over whether or not time could be a substance. This shift follows from Descartes' definition of substance in *Principles I.51* as being "a thing which exists in such a way as to depend on no other thing for its existence."[1] Descartes takes the capacity for independent existence to be the criterion that establishes something to be a substance; those things *in*capable of this sort of independence are thereby established to be mere attributes of substance. Thus, within the Cartesian metaphysical framework, the historical debate between the absolutists and reductivists translates into a debate over whether time might exist as a Cartesian substance or instead as an attribute possessed by substances. Since Descartes' predecessors and contemporaries were actively debating over absolutist vs. reductive conceptions of time, this approach considers Descartes' account time as it might have been considered by Descartes' own contemporaries.

ONTOLOGICAL CATEGORIES IN DESCARTES

Though Descartes defines substance as being "a thing which exists in such a way as to depend on no other thing for its existence,"[2] it is important to note that Descartes also claims that only God achieves this true degree of independence. Though God is the only fully independent substance, Descartes asserts that the term "substance" can be applied to things that only need "the ordinary concurrence of God to exist." Such entities are appropriately called 'substances' since they are distinguished from those things that "cannot exist without other things" (i.e., qualities, attributes, or modes).[3] Thus, in wondering if time might exist as an independent entity/substance, I do not wonder if time might be on an ontological par with God. Rather, following Descartes' own usage, I wonder if time has the sort of independent existence enjoyed by other created substances. That is to say, I am questioning whether time has the status of being a thing that depends on nothing beyond God's concurrence for its existence.

As Descartes proceeds in his discussion between *Principles I.51* and *I.52*, he emphasizes the importance of determining a thing's level of dependence in order to establish its ontological status. After defining substance (in both its primary and secondary sense), Descartes proceeds to discuss attributes. Just as he identifies substances by their *in*dependence, he identifies attributes by their *de*pendence. Descartes explains, "if we perceive the presence of some attribute, we can infer that there must be present an existing thing or substance to which it may be attributed."[4] Whenever an attribute is perceived, a substance can be inferred, because attributes depend on the existence of the substance of which they are an attribute. In other words, Descartes makes it clear that his metaphysics prohibits "free-floating" attributes.

It is worth noting that Descartes does make a further distinction within the realm of dependent things. Among dependent things, Descartes distinguishes (1) principal attributes which "constitute [a substance's] nature and essence" and (2) 'modes.' He identifies extension as the principal attribute of corporeal substance, and thought as the principal attribute of thinking substance. Though modes are also attributable to a substance, they differ from thought and extension insofar as modes are non-essential ways that a substance can be. In noting Descartes' distinction between principal attributes and modes it is important to emphasize that this distinction does not hinder Descartes' ontology from offering a clean break between those things that depend on other created things, and those that do not. What the distinction between principal attributes and modes indicates is merely that Descartes offers two different ways for this *de*pendence to be manifested.

Whereas neither principal attributes nor modes can exist without a substance, one can think of modes as being additionally "dependent" on a principal attribute since modes are ways for a substance to be either extended or thinking. As Garber explains, "the term 'mode,' *modus* in Latin, means quite simply "way"; it is a perfect term to choose to express the fact that all accidents are ways of something being extended or being a thinking thing, *ways* of expressing the essence of a thing."[5] Thus, modes and principal attributes are both dependent on substances, but whereas attributes are ways that substances are in general, modes are ways that either extended or thinking substances are in particular. Though there are important differences in how modes and principal attributes are manifested, these differences will not be further explored in this context. Such an exploration is not required, since Descartes' ontology cleanly distinguishes between things that depend for their existence on other created things, and those that do not. For my purposes, I will call the former "attributes" and the latter "substances."

WHY THINK THAT TIME COULD BE A SUBSTANCE FOR DESCARTES?

Though scholars seem sometimes to discuss Cartesian time as though it is a thing that can be studied as an isolated entity, such an approach does not focus on the question of whether such independence would entail the substantiality of time. To understand why one might think that Cartesian time is the sort of thing that *can* be considered in isolation of other things, one need only look to the primary texts that have inspired most discussions among Cartesian commentators. Among Descartes' discussions on time, one of the most significant, and certainly most discussed, passages is the discussion that he offers in the second causal argument of the third *Meditation*—an argument that also appears in *Principles I.21*. In *I.21*, Descartes claims that "the nature of time is such that its parts are not mutually dependent, and never coexist."[6] Given this fact about time, Descartes argues that the duration of one's existence offers sufficient grounds for proving God's existence. In brief, Descartes argues that since the parts of time are independent from each other (and thereby causally separated)[7] something's existence at one moment cannot be the cause of its existence at the next moment. Since beings endure through multiple moments, and since this continued enduring cannot be credited to the thing's existence at a prior moment, Descartes argues that this enduring must be caused by the immediate concurrence of God. Thus, Descartes bases a proof for God's existence on the "parts" of time and on the fact of one's continuing duration, since these two facts together necessitate the existence of God as the immediate cause for this continuing duration.

Descartes' proof of God's existence based on the nature of time appears in *Principles I.21*, in *Med III*, and in his *Replies* related to *Med III*. Many commentators have focused on the account of time suggested by these passages because it seems surprisingly atomistic. That is to say, many have looked at these passages and wondered if Descartes' parts of time doctrine suggests the existence of discrete moments of time that are either continuously or discontinuously linked. Despite the broad interest in the atomism question, this is not the topic of interest for this chapter.[8] Rather, the parts of time doctrine is here important because it explains why some might be tempted to read Cartesian time as being a type of independent entity—and so, as a type of Cartesian substance.

In *Principles I.21*'s claim that "time is such that its parts are not mutually dependent,"[9] the description suggests that time is a composite entity made up of distinct parts. Given that Descartes' ontology divides between substances and attributes, and that attributes are a *way* that a substance is, by Descartes describing time as a composite of parts, it is difficult to see how such a thing might be a *way* that something is.[10] If Descartes had simply described time to be orderly divisions, then such an account would be consistent with its being an attribute of substance. To the question, "In what way is the enduring of my lifespan?" the reply would be "orderly divided." Since Descartes does not identify time with orderly divisions, however, but instead as being a whole composed of independent parts, it doesn't make sense to claim that this whole is some kind of way that something can be. Are *ways that things are* made up of discrete parts? It is hard to understand what a composite *way of being* might be like. By reflecting on the strangeness of this idea, one might suppose that time is not some kind of attribute, but rather a kind of independent entity (i.e., a substance).

Given the strangeness of there being *ways of being* that are made up of distinct parts, it is not surprising that some commentators who focus on the parts of time doctrine (perhaps, inadvertently) seem to credit Descartes with offering a substance-like account of time. This is a problematic tendency insofar as scholarship related to Descartes' account of time has primarily focused on the nature of these temporal parts. This has been the focus because analysis has generally related to determining whether the parts in Descartes' account of time are continuously or discontinuously linked. In *Descartes' Metaphysical Physics*, Daniel Garber provides a brief overview of this classic debate between proponents of the discontinuity and continuity theses. Garber explains that the former, "received" view proposes that Descartes presents time as being composed of discontinuous "temporal atoms." [11] As such, the received view clearly follows the language of "independent parts" sketched above. Accordingly, this type of interpretation is particularly amenable to regarding Descartes as positing time to be an independent entity that exists apart from those things that are in time.

One can see how a proponent of discontinuous temporal atomism might credit Descartes with offering a substance-like account of time in Ken Levy's recent (2005) article, "Is Descartes a Temporal Atomist?" [12] In this article, Levy argues that Descartes offered a strongly discontinuous [13] variety of temporal atomism. He explains his interpretation by using the analogy of a cut string. He writes:

> [Some philosophers] become susceptible to one or another version of temporal atomism when they treat time as though it were a piece of string that can be cut into separated parts and then mistakenly take this model to be an accurate representation of time. Descartes, I shall argue, is just one such philosopher. [14]

In other words, Levy argues that Descartes held time to be something that exists as a composite entity composed of disconnected temporal atoms (which are conceived of as really distinct parts). As a participant in the atomism debates, Levy's argument (like the arguments of many others) does not begin by inquiring about the ontological status of the string, but rather by focusing on its internal structure.

In conceiving of Descartes' account of time as a composite whole (i.e., as a string with parts), Levy shows a tendency to treat time as if it were only dependent on its "parts" and not dependent on some substance—as would be the case if time were an attribute. Levy argues, for example, that the question of whether one moment of time will be followed by another can be treated independently from the question of whether a substance "in" time will continue to endure. He writes:

> If it is logically possible that time end at any given moment *t*, then it is certainly logically possible that I—who exist in time—cease to exist at *t*. But it is not clear that [my not being guaranteed to exist at a next moment] depends on [the fact that the parts of time are independent of each other]. . . Even if every earlier part of time logically guaranteed that there be a later part of time, if it were (therefore) logically impossible that time end, it would still be the case that I could cease to exist at any moment. Time's immortality would hardly guarantee my immortality. So the fact that my ceasing to exist is logically possible does not necessarily depend on whether or not it is logically possible for time to end. [15]

In this argument, if one were to conceive of time as an attribute of a thing in time, then Levy would be arguing that the existence of the substance is not dependent on the existence of the attribute. Though it is of course correct that attributes depend on substances and not vice versa for Descartes, by divorcing the question of time's continuation from the question of the continuation of a thing in time, Levy shows his tendency to presume that the existence of something in time is irrelevant to the question of time's existence. This tendency treats Cartesian time as if it were a Newtonian container. But, if time might be a type of attribute, then such a divorce is a mistaken way to conceive of Cartesian time. Though it would still be fair for Levy to claim that an infinite time would not guarantee the immortality of any particular substance, such a determination would guarantee the eternal existence of some substance or other. Since there can be no free-floating attributes, if logic demanded time's eternal existence, then an eternal time would also demand the eternal existence of some substance of which time is an attribute.

Levy elsewhere makes his presupposition that time is a substance more explicit. Indeed, Levy goes so far as to argue against regarding time as a type of attribute. He does this by arguing that time is something other than duration, and that only duration is an attribute that inheres in a substance. He claims, "it does not really make sense to say that time itself is 'in the very things of which [it is] said to be [an] attribute or [mode].'. . . we tend to think not that time is in (no less an attribute of) all things but rather just the opposite – that all things are in time."[16] Accordingly, Levy concludes that "time is not equivalent to 'duration taken in the general sense' and therefore is not necessarily '[an] attribute or mode.'"[17] These passages clearly indicate Levy's belief that Descartes offers some account of time that is not merely an attribute of substance. Levy argues that time in Descartes is a thing *in which* temporal things endure (like the Newtonian container view), and not that time is something dependent on *its* enduring *within* temporal things.

Though Levy's account only shows one example of someone attributing a sort of independent existence to Cartesian time, reflecting on the "parts of time" language has shown why such a tendency might be reasonable. Such a tendency carries significant metaphysical weight, however, since it further assumes that time is a type of Cartesian substance and thus that Descartes is presupposing an absolutist account of time. Having looked at the textual evidence that seems best to support an absolutist reading of Descartes, I can now consider the significant evidence against such an interpretation. The contrary evidence proves that any tendency to presume a sort of absolutism in Descartes' account of time is not only mistaken, but one that establishes a significant road block to understanding the nature of Cartesian time.

FOUR ARGUMENTS AGAINST TIME BEING A CARTESIAN SUBSTANCE

Though Descartes' "parts of time doctrine" may seem to invite the interpretation that time is a sort of composite substance, Descartes provides other discussions that exclude this interpretation. By considering these discussions, I will provide four arguments against time's being a Cartesian substance. I will refer to these arguments as:

1. The Argument from Attributes
2. The Argument from *Principles I.57*
3. The Argument from Radical Contingency
4. The Argument from Numerically Distinct Parts.

Whereas arguments (1) and (2) will focus on various metaphysical commitments that Descartes offers in the *Principles,* arguments (3) and (4) will offer related arguments that focus on Descartes' proof for God's existence based on the (troublesome) "parts" of time. I will conclude that contrary to first appearances, Descartes' "parts of time doctrine" does not entail the substantiality of time. Rather, this doctrine itself supports an attribute theory of time.

The Argument from Attributes

The first textual argument against time's being a Cartesian substance can be extracted from *Principles I.52* and *I.53*. This 'argument from attributes' follows Descartes' claim at *I.52* that "corporeal substance and mind" are the only two types of created substances. This claim, when conjoined with *I.53*'s claim that there is just one primary attribute characteristic of each of these types of substances (i.e., extension for body and thought for mind) provides a test for determining if something is a created substance; namely, if X is a created substance, then X is either a corporeal substance or a mind. If X is a corporeal substance, then it is extended in "length, breadth and depth" and if X is a mind, then thought constitutes its nature. Thus, for time to be a Cartesian substance, it would need to be either spatially extended or thinking. Yet, both of these descriptions seem inapplicable to an account of time. Indeed, it is difficult to conceive how any account of time could credit time itself with being either spatially extended or thinking. Thus, unless a proponent of the substance interpretation could show how the substantial time it defines could be either spatially extended or thinking, then this account is inconsistent with the metaphysical landscape that Descartes defined.[18]

The Argument from Principles I.57

A second textual argument against time's being a substance is what I refer to as 'the argument from *Principles I.57*.' In *I.57* Descartes explicitly states his view that time is a type of attribute. This passage clearly offers an argument against time's being a substance so long as one grants that Descartes offers a consistent account of time between this principle and the "parts" language of the *Meditations* and elsewhere. In *Principle I.57*, Descartes addresses the question of what types of beings time and duration are, given that, "some attributes are in things and others in thought."[19] Since *I.57*'s stated agenda is merely to determine the *type* of attributes that duration and time are, this passage shows that Descartes is not even considering the possibility that time might be a sort of substance. That time is some sort of attribute is a fact that he simply takes for granted. Given that *I.57* simply assumes that time is a type of attribute, and since Descartes could not consistently maintain that time is both a type of attribute and a type of substance, this passage provides strong evidence against reading the "parts" language of *Med III* as positing time to be a type of aggregate substance.

Though it may seem as though *I.57* is alone sufficient for refuting the hypothesis that time is a Cartesian substance, one might doubt the conclusiveness of this passage by arguing that *I.57* is so unlike the "parts" account of time that one ought to simply assume that Descartes offered inconsistent views on time. Indeed, when Ken Levy briefly discusses *Principles I.57*,[20] he suggests that the apparent inconsistency between the parts account of time and the account offered in *I.57* is significant enough to warrant treating this passage as a mere anomaly. Indeed, he claims that one can infer very little from "minor passages tucked away in the *Principles*."[21] Even if one disagrees with how Levy's evaluates the significance of *I.57*, one might still agree that this passage is so unlike the discussions of time given in the *Med III* proof that it is appropriate to treat it as anomalous.

It is certainly reasonable to suppose that Descartes' views on time might have undergone changes, or even to think that his views might not always have been consistent. In the following two arguments, however, I not only provide additional grounds for denying that time is a Cartesian substance, I also show why *I.57* should not be read as anomalous. That is to say, I show how a close examination of Descartes' claims about the parts of time indicates why this "parts" account is itself inconsistent with time's being a substance.

The Argument from Radical Contingency

For my third argument against time's being a Cartesian substance, I focus on the proof for God's existence based on the parts of time. Remember that this argument *seems* to presuppose the independence (and thereby the substantiality) of time. Contrary to first impressions, this proof actually presupposes the *impossibility* of the substantiality of time. In the following 'argument from radical contingency,' I consider what is required for the possibility of individual moments. Recall that in the *Med III* proof Descartes explains that a "lifespan" can be divided into "countless parts, each completely independent of the others."[22] From this, Descartes concludes, "it does not follow from the fact that I existed a little while ago that I must exist now, unless there is some cause which as it were creates me afresh at this moment—that is, which preserves me."[23] In other words, the *Med III* proof establishes the radical contingency of substance. Descartes establishes that the existence of every substance is contingent on God's concurrence (i.e., creation/preservation) at each and every moment as the immediate cause of its continued existence.

By focusing on the radical contingency thesis, one learns something about the nature of created substances and how they are dependent on God. In *Principles I.51,* when Descartes defines substance he explains that created substances are not substances in the same way that God is a substance because "they can exist only with the help of God's concurrence."[24] The analysis of the parts of a lifespan given in *Med III* clarifies the nature of this concurrence. Descartes argued that created substances are substances insofar as they depend on *nothing other than* the concurrence of God. As the radical contingency thesis shows the way in which a substance is dependent on God, it provides a standard for determining if something within the Cartesian metaphysic might be a substance; namely, does the thing in question possess the sort of independence whereby it could exist at successive moments given divine concurrence *alone*? In the following argument, I will show why time cannot meet this necessary condition for substantiality:

The Argument from Radical Contingency

1. Time has parts (namely, moments). (*Med. III*)
2. According to radical contingency, substances can exist at one moment (that is, could exist even if there were no prior or later moments of them).
3. Suppose time is a substance. (for *reductio*)
4. If time is a substance, then time can exist at one moment. (2, 3)
5. Thus, each moment of time could exist even if there were no prior or later moments of time. (The most charitable interpretation of the consequent of 4)
6. Moments of time are essentially relational.
7. Anything that is essentially relational requires the existence of at least two *relata* for its possibility (namely, itself and that to which it is related). (Definition of 'essentially related')[25]

8. It is not possible for a moment of time to exist if there are no prior or later moments of it. (6, 7)
9. Lines 5 and 8 contradict.
10. Therefore assumption 3 is false. Time is not a substance.

There are certain oddities in this argument that cannot be avoided, in addition to some controversial premises. It is undoubtedly odd, for example, to try to conceive of time existing at a moment (i.e., line 4). Though it may indeed seem bizarre to conceive of time *at a time*, the oddness in this maneuver is compelled by the analysis of radical contingency just considered. If my analysis is correct, then Descartes' radical contingency doctrine provides a test for all substances; namely, if X is a substance, then X can exist at a moment. Thus, to analyze the hypothesis that time itself might be a substance, I plug time in for X, and end up with line 4 of the argument. As I indicate in line 5, it seems that the most charitable way of applying this general rule to time is to take it as suggesting that if time were a substance then it would be possible for one moment of time to exist in isolation of other moments. Accordingly, as the argument proceeds, I consider whether it would be possible for a substantial time to exist at a single moment, and determine that the answer is no.

The most controversial premise of this argument is surely found on line 6, where I claim that "moments of time are essentially relational." This argument attempts to construct the most reasonable position that someone positing time to be a Cartesian substance might offer. If one is arguing that time is a substance in Descartes, then one is explicitly arguing that time is capable of independent existence. To argue that time is capable of independent existence, however, one must be able to offer some account of what it would be for time to exist at all. Given that Descartes posits time to be a collection of parts (i.e., moments), one must be able to give some account of what it is for a moment to exist as such. In my argument's assertion that moments are essentially relational, I merely suggest that what it is for a moment t_2 to exist is for it to be a moment subsequent to moment t_1 and prior to moment t_3. In other words, I argue that t_2's nature is "essentially relational." As I explain in line 7, "anything that is essentially relational requires the existence of at least two *relata* for its possibility (namely, itself and that to which it is related)." With this seemingly plausible account of moments in hand, I then simply consider the criteria for substance that are provided by Descartes' account of radical contingency in conjunction with this account of how moments might be individuated. The *Argument from Radical Contingency* shows that time could not both meet the criteria for substance suggested by his account of radical contingency and maintain a relationalist account of moments. Thus, time could not be a Cartesian substance, because it would not possess the sort of independence whereby it could exist at a single moment given divine concurrence alone. Time could not exist "at a moment" by divine concurrence alone because each moment of time seems minimally to require the existence of an additional moment to which that moment could be related.

The Argument from Numerically Distinct Parts

Though I believe that the *Argument from Radical Contingency* assumes a reasonable account of moments and then proves that time could not be a type of Cartesian substance, it seems that one might object to the important premise that moments are "essentially relational" on textual grounds. One might claim that this particular account of moments (despite its plausibility) could not have been held by Descartes, since in *Principles I.21*, Descartes claims that the "parts [of time] are not mutually dependent, and never coexist."[26] Since *Principles I.21* denies that moments are mutually dependent, it seems that *I.21* might show that Descartes cannot appeal to the external relations between moments as establishing the nature of an individual moment. He could not appeal to these external relations since this would make the nature of individual moments mutually dependent. In other words, the very idea behind the *Argument from Radical Contingency* (namely, that time could not be a substance because no moment can exist in isolation) appears to be denied by *Principles I.21*. This principle seems to suggest that the nature of a moment cannot be determined in virtue of its location in the timeline, since this hypothesis makes moments mutually dependent and Descartes explicitly states that moments are not mutually dependent.

Though this interpretation of *Principles I.21* may indeed pose a significant difficulty for line (6) in the *Argument from Radical Contingency*,[27] this worry actually invites another, simpler argument against reading Descartes' account of time as being consistent with its being a substance. Namely, it invites the 'argument from numerically distinct parts.' In this, my fourth and final argument, I argue against time's being a Cartesian substance while assuming that moments are not essentially relational.

We have just seen that *Principles I.21* offers one of Descartes' strongest statements of his parts of time doctrine by its assertion that "the nature of time is such that its parts are not mutually dependent, and never coexist."[28] With the parts of time doctrine in hand, an immediate problem with any account positing that Descartes held time to be a substance presents itself. If Descartes maintained that time is composed of independent, separable parts, then one can ask how these parts could be distinguished if time were a substance? If one denies that moments can be distinguished in virtue of the external relations between moments, then one must appeal to some other basis for the distinction between moments. There must be some basis or other for a distinction between moments, since a minimum criterion for something's existing at all would seem to be that there is something that it is to be that thing. In other words, for a thing to exist, there must be something that gives that thing a distinct nature. If one excludes an appeal to the essential relations between moments (based on *Principles I.21*), then it seems that there are very few other alternatives capable of providing moments with a distinct nature.[29]

To claim that Cartesian time is a substance one must grant that time is capable of independent existence. Thus, a proponent of the substance thesis cannot claim that moments of time are individuated by appealing to any other thing, including things that are "in time." That is to say, the substance thesis excludes one from claiming, for example, that t_1 is distinguished from t_2 by way of the occurrences happening at t_1 or t_2. These occurrences could not ground a distinction between t_1 and t_2 because this would make it the case that the identity of t_1 and t_2 depends on those occurrences. To posit that the identity of t_1 and t_2 depends on the occurrences happening within them would be to contradict the claim that time exists as an independent entity. Though the thesis that moments are individuated by external relations between moments offered the advantage of making time only dependent on itself (i.e., only dependent on its own various parts), if *Principles I.21* disproves this option (at least on this reading) then it seems that all reasonable grounds for a distinction between moments are lost. If the distinction cannot be grounded on (1) occurrences in time, or on (2) relations between moments, then it seems the only other alternative is (3), that moments are distinguished in virtue of individual essences or *haecceities* possessed by each moment.[30] This last suggestion is, however, difficult to take seriously since it is hard to even conceive what such an individual essence for a moment might be, and since there is no textual evidence to indicate that Descartes held such a view. If the sheer implausibility of (3) is sufficient for excluding it, then it seems a simple argument against time's being a substance presents itself. For, if it is true that there would be no basis in the world for a distinction between these parts (i.e., moments) of time, then it seems that time would not possess the numerically distinct parts that Descartes posits for it. More formally, my argument is as follows:

Argument from Numerically Distinct Parts

1. Time has numerically distinct parts (Premise 1, *Principles I.21*)
2. If time were a Cartesian substance, then it could not have numerically distinct parts (because these parts could not be individuated).[31] (Premise 2, argued above)
3. Thus, time is not a Cartesian substance (by *Modus Tollens*).

In this argument, Premise 2 is obviously doing all of the work. This premise was defended above. In short, I argued that moments could not be distinguished by

occurrences happening at a time because then time would lack the independence that the substance theorist wants to attribute to it.
relations between moments because that would make moments essentially relational (thus subjecting the defender of this claim to the Argument from Radical Contingency).
or *their individual essences* because such a view is a highly implausible interpretation of Descartes.

Thus, this final argument shows the inconsistency in positing that time could both be a Cartesian substance and that it could be composed of parts. As the very possibility of time's possessing numerically distinct parts demands that there exist some grounds for a distinction between parts, and since a substantial time would not admit of any of these grounds, the two views contradict. It seems that Descartes *could not* consistently hold that time is composed of parts and hold that it is a substance. Thus, even when focusing on Descartes' parts of time doctrine, one finds strong reasons for denying that time could be a substance.

CONCLUSION

Having considered evidence for and against the view that Descartes might have granted time the status of a substance, I have shown that the weight of evidence opposes this position. Through the four arguments of section III, I have shown that there are both strong textual and philosophical grounds for denying that Descartes could have taken time to be a type of substance. By showing the inconsistencies that would hamper such a position, I have argued that there are deep metaphysical reasons that could have prompted Descartes' claims in *Principles I.57*. Given that the grounds explaining *I.57* are extracted from an analysis of the parts of time doctrine, I have thereby additionally shown strongly that *I.57* should not be treated as an anomalous passage. I have shown that Descartes' discussions of the parts of time reinforce the importance of *I.57* insofar as this principle explicitly states what is only entailed by other discussions; namely, that time is an attribute and not a substance. When *I.57* is conjoined with the analysis of the parts of time doctrine, one discovers that time not only *is not* but indeed *could not be* a Cartesian substance.

NOTES

1. *Principles I.51*: AT VIIIA 24, CSM I.210.
2. *Principles I.51*: AT VIIIA 24, CSM I.210.
3. *Principles I.51*: AT VIIIA 24, CSM I.210.
4. *Principles I.52*: AT VIIIA 25, CSM I.210.

5. Daniel Garber, *Descartes' Metaphysical Physics* (Chicago: The University of Chicago Press, 1992), 69.

6. AT IXB 13, CSM I.200.

7. For a very good discussion on this point, see Geoffrey Gorham, "Cartesian Causation: Continuous, Instantaneous, Overdetermined," *Journal of the History of Philosophy* 42 (2004): 389–423.

8. This discussion is the focus of chapter 5.

9. AT IXB 13, CSM I.200.

10. Though Descartes of course distinguishes in *Principles I.56* among modes, qualities, and attributes, this still seems to be a fair classification, since modes and qualities are also ways that a substance can be, though they are ways that would cause the substance to be modified or distinguished.

11. Garber, *Descartes' Metaphysical Physics*, 266–268.

12. Ken Levy, "Is Descartes a Temporal Atomist?" *British Journal for the History of Philosophy* 13 (2005): 627–674.

13. "Strongly discontinuous" is the view that different moments of time are not only separate, but actually separated by a gap in which there is no time at all. See Levy, "Is Descartes a Temporal Atomist?" 633.

14. Ibid., 628.

15. Ibid., 637.

16. Ibid., 663.

17. Ibid., 663.

18. Given contemporary discussions on spacetime one might want to argue that Cartesian time could be spatially extended. However, this position seems to offer a highly implausible interpretation of Descartes' own view. As Descartes had so much to say on the nature of space and extension, and had so little to say on the nature of time, it seems evident that these categories were not conjoined in his own thinking.

19. AT VIIIA 26; CSM I.212.

20. Levy, "Is Descartes a Temporal Atomist," 662.

21. Ibid., 662. I quite strongly disagree with Levy's characterization of *I.57* as a "minor passage" or even the suggestion that anything offered in the *Principles* might count as "tucked away."

22. AT VII 49; CSM II.33.

23. AT VII 49; CSM II.33.

24. AT VIIIA 24; CSM I.210.

25. In this definition, I set aside the case of an identity relation, since this is not a distinguishing type of relation.

26. AT IXB 13, CSM I.200.

27. It is worth noting, however, that this type of "worry" only exists for someone who is reading moments as existing as parts of an independent, substantial time. If, instead, one reads moments as stages of an enduring thing (which I think is correct), then there is not a problem of how to individuate moments, since moments would be individuated by the substance of which they are an attribute, and would thus not need a ground for their independent existence as such.

28. AT IXB 13, CSM I.200.

29. Bonnen and Flage note a similar concern when they describe the difficulty of trying to conceive of "bare atoms" of time. Clarence Bonnen and Daniel E. Flage, "Descartes: The Matter of Time," *International Studies in Philosophy* 32 (2000): 3.

30. Alexander U. Bertland of Niagara University suggested a fourth option for how individual moments might be distinguished for Descartes. As his suggestion seems both clever and initially plausible, it is worth considering. Dr. Bertland wondered if individual moments might be distinguished just in virtue of their unique creation acts by God, that is, moment t_1 is distinguished from moment t_2 in virtue of the fact that moment t_1 was created prior to moment t_2. Though this suggestion seems initially appealing, there are a couple of significant flaws with this option. First, if the differences between moments are caused by differences in particular creative acts, then this

would suggest a violation of Descartes' commitment to divine immutability. For different moments to be differentiated by their different causal acts alone, then each causal act would need to be fundamentally different in a way that the acts of a radically simple and unchanging being could not seem to be.

31. Though one might object that this premise confuses an epistemological fact with an ontological one, I think this objection misses the idea. I am not arguing that one could not distinguish one moment from another without there being some grounds for distinction (though of course this would be true). Rather, I am arguing that the very possibility of there being distinct parts requires that there be some basis upon which an actual distinction might rest. That is to say, some basis upon which it is true to say that moment t_1 is not identical to moment t_2.

THREE

Descartes' Temporal Dualism

In the first chapter I noted that proponents of a reductive approach generally claimed that time is dependent on motion—either through its being reducible to a particular motion (i.e., celestial reductionism) or through its being reducible to a relation between motions (i.e., relationalism). In the second chapter I explained *why* time cannot be a Cartesian substance, and thus why Descartes must favor the reductive approach over the absolutist approach. I can now consider what kind of reductive approach Descartes favors by considering what role motion plays in Descartes' account of time.

Since Descartes' texts so frequently link discussions of time with discussions of motion, it is evident that Descartes at least followed the historical precedent of conceiving of one in terms of the other. However, Descartes' account diverges from the traditional, reductive approaches by seeming to offer two, inconsistent ways of relating time and motion. If Descartes were following a simple reductive approach, then he would need to maintain that time is ontologically dependent on motion. Though Descartes does appear to presume this in some passages, he elsewhere seems to suggest the opposite; namely, that *motion* depends on time.

Rather than following one of the more common reductivist approaches, I here argue that Descartes follows a dualist approach to time. I argue for this claim by analyzing the seemingly inconsistent ways that Descartes relates motion and time. As temporal dualism offers a clear resolution to these inconsistencies, while also being consistent with Descartes' own texts, there is strong support for interpreting Descartes as a temporal dualist. In claiming that Descartes is a temporal dualist, I am claiming that he posits two really distinct temporal attributes. The first kind (which I call 'duration') *is not* dependent upon motion for its existence. The second kind (which I call 'time-in-thought') *is* somehow dependent upon motion.

This chapter has two major goals—first, to establish the fact of Descartes' temporal dualism; second, to begin sketching the character of this dualism. The first goal is addressed in sections I and II. In section I, I identify the various, apparently inconsistent relationships that Descartes suggests between time and motion in several significant texts. In section II, I show how *Principles I.57* shows Descartes offering a real distinction between the two temporal attributes: duration and time-in-thought. As this real distinction suggests temporal dualism, and this dualism offers a resolution to the apparently inconstant accounts of time suggested in other passages, *Principles I.57* does the major work for establishing my dualist interpretation. Having argued for the fact of Descartes' temporal dualism, I then (in sections III and IV) begin analyzing the nature of this dualism. In section III, I begin considering each of the two attributes in isolation from each other. In section IV, I compare Suarez's dualist account of time with the view emerging of Descartes' account. This final task continues into chapter 4, since noting how Descartes' view diverges from that of Suarez helps show the interesting and unusual character of Descartes' temporal dualism.

RELATING TIME AND MOTION

Though reductivists had good philosophical grounds for supposing that time is dependent on motion, there are also good philosophical reasons for thinking instead that motion is dependent on time. In different passages, Descartes seems to reference and endorse the grounds supporting each of these contrary dependence relations. For example, in *Principles II.39*, Descartes seems to point to reasons why one would claim that motion depends on time. In this text, he claims that "no motion takes place at a single instant."[1] In asserting that there is no motion in a single instant, Descartes appears to acknowledge the uncontroversial point that there can be no motion or change without some type of temporal succession, since motion is itself inherently successive. That is to say, since something cannot simultaneously hold contrary properties, the very possibility of movement or change in something presupposes that the thing is able to be one way at t_1 and another way at t_2.

Though *Principles II.39* suggests that Descartes conceives of motion as dependent on time, he elsewhere seems to endorse the reductivist stance that time is instead dependent on motion. There are additional, good philosophical grounds why Descartes might have maintained this contrary order of dependence. The central reason why Descartes, like others before him, would plausibly suppose that time depends on motion is because time is inconceivable without motion. Without some kind of motion or change, two moments or states of an individual substance are indistinguishable. As any *conception* of time requires that successive moments are distinguishable according to before and after, motion is required to make time conceivable. Descartes discusses this point in *Rule 12*, when he explains that the concepts of duration and time are necessarily conjoined with a concept of motion because "we cannot conceive either of them distinctly if we judge them to be separate from [motion]."[2]

Though *Rule 12* only claims that motion is required for time to be *conceived*, this epistemic point can have ontological import for anyone who endorses the reductivist suggestion that time is a "measure of motion." This Aristotelian phrase is one that Descartes explicitly employs in *Principles I.57*, when defining time "as distinguished from duration."[3] This phrase and the view it suggests are quite significant since, if time is the measure of motion, then the very possibility of time existing would thus be dependent on time's being conceivable by a mind. As Aristotle himself claimed, "if there cannot be someone to count there cannot be anything that can be counted either."[4] In other words, if time cannot be *conceived* without motion, then it cannot *exist* as a measure, since measures require a measur*er*. Thus, if Descartes held time to be the measure of motion, then the fact that time could not be conceived without motion would entail that time could not exist without motion. In this case, an epistemic fact about what is required for time to be conceived carries ontological import. Accordingly, it seems that there are good philosophical grounds that might have compelled Descartes to think that time depends on motion, just as there were good philosophical grounds for him to think that motion depends on time.

If Descartes had good reasons for thinking that motion is ontologically dependent upon time *and* for thinking that time is ontologically dependent upon motion, then one might suspect that Descartes will suggest that time and motion are co-dependent (i.e., to have one you must have both). If, however, one returns to *Principles I.57*, it seems that Descartes rejects this option. In *I.57*, Descartes argues that duration (a term that he often uses interchangeably with 'time') does *not* depend upon motion for its existence. He argues that the duration of things in motion is the same as the duration of things which are not moving.[5] In claiming that the same duration obtains regardless of motion, Descartes implies that time (or at least duration if there is a distinction) must be ontologically *in*dependent of motion. Thus, in *I.57*, Descartes appears to reject the co-dependency of motion and time. But if he rejects the co-dependency option, then he seems committed to a contradiction—that time both is and is not dependent upon motion for its existence. To determine if Descartes might really hold contradictory views, it seems that one must return to the various passages where he relates the two.

Recall the relationship between time and motion that Descartes gives in *Rule 12*. In this text, Descartes relates motion with time and duration while explaining the types of conjunction that can exist between simple things.

> The conjunction is necessary when one of them is somehow implied (albeit confusedly) in the concept of the other so that we cannot conceive either of them distinctly if we judge them to be separate from each other. It is in this way that shape is conjoined with extension, motion with duration or time, etc. because we cannot conceive of shape which is completely lacking in extension, or a motion wholly lacking in duration.[6]

There are several important points to note in this passage. As already noted, this passage is explicitly discussing an epistemic point (i.e., the requirements for a thing to be conceived). But it seems that Descartes is intentionally making a metaphysical point when explaining what things are *necessarily* conjoined. For example, Descartes explains that the concept of shape is necessarily conjoined with the concept of extension. The explanation for this epistemic point is a metaphysical fact. One cannot conceive of a shape that lacks extension because shape is a mode of extension—a way that extension is. Without there being a thing that is extended in some way or other, there is no shape. Accordingly, one cannot conceive of shape without extension because this would mean that one was trying to conceive of something that is extended in some way which lacks extension—an impossible being. That we cannot conceive of impossible beings is not a novel claim for Descartes to be making—just as we cannot conceive of square triangles, we cannot conceive of unextended bits of extension. The epistemic limit in these cases is the natural consequent of the metaphysical impossibility of the thing we are trying to conceive. This point is worth noting since it suggests that the other epistemic limit *Rule 12* identifies (i.e., that the ideas of duration or time and motion are necessarily conjoined) is one that also makes a metaphysical point and not merely an epistemic one; namely, that time or duration are conceptually dependent on motion *because* they are metaphysically dependent on motion.

Another interpretive point to note in *Rule 12* is that is unclear how the 'or' separating duration and time should be interpreted.[7] One might suppose (a) that the 'or' distinguishes the duration and time as two different things, both of which are conjoined with motion in the same way, or one might suppose (b) that the 'or' indicates that Descartes is speaking loosely and that he uses the terms 'duration' and 'time' as equivalent ways of picking out the same thing.[8] As (a) and (b) seem to offer equally reasonable interpretations of *Rule 12*, this passage must be interpreted in light of both possibilities. In other words, this text might be interpreted as claiming either (a) that time and duration are each distinctly dependent on motion, or (b) that time/duration (with each term loosely connoting the same object) depends on motion.

Recall also how Descartes relates time and motion in *Principles II.39*. In *II.39*, Descartes claims that "no motion takes place in a single instant of time."[9] In other words, *II.39* suggests the intuitive point that the very possibility of motion requires a succession of moments to ground the possibility of change/motion. The intuitive reasons for supposing that motion depends on some sort of temporal succession have already been discussed. It is worth noting, however, that when Descartes here suggests that motion requires time, he does not thereby exclude time from also being dependent on motion. In claiming that there is no motion at a single instant, Descartes clearly suggests more about the nature of motion than about the nature of time. Accordingly, this passage does not exclude Descartes from holding that some sort of two-way dependence could exist between time and motion. Though it is obviously not the case that there could be a two-way dependence between motion and time if each dependence required the chronological priority of the one before the other,[10] there could be a two-way dependence if motion and time could be jointly and simultaneously dependent—one on the other. Or, a type of two-way dependence could be possible if (as I argue), "time" is a general term that picks out two ontologically distinct entities.

THE REAL DISTINCTION IN *PRINCIPLES I.57*

As *Principles 1.57* is the most significant passage where Descartes relates time and motion, I will present this passage separately and with particular attention. To do this, I must first address a few interpretive puzzles, and then show how the text explains Descartes' route for making motion both dependent on time and required for time without assuming a contradiction. Below is the text in its entirety.

> Principles 1.57:
> Now some attributes or modes are in the very things of which they are said to be attributes or modes, while others are only in our thought. For example, when time is distinguished from duration taken in the general sense and called the measure of movement, it is simply a mode of thought. For the duration which we understand to be involved in movement is certainly no different from the duration involved in things which do not move. This is clear from the fact that if there are two bodies moving for an hour, one slowly and the other quickly, we do not reckon the amount of time to be greater in the latter case than the former, even thought the amount of movement may be much greater. But in order to measure the duration of all things, we compare their duration with the duration of the greatest and most regular motions which give rise to years and days, and we call this duration 'time'. Yet nothing is thereby added to duration, taken in its general sense, except for a mode of thought.[11]

SOME PRELIMINARY POINTS ABOUT INTERPRETING *PRINCIPLES I.57*

One immediate point to note in this passage is that Descartes claims, "when time is distinguished from duration taken in the general sense and called the measure of movement, it is simply a mode of thought." As Descartes' phrasing indicates that he is picking out a special sense of the term 'time' (namely, the sense that can be distinguished from duration) it seems that one should not assume that this sense refers to Descartes' primary notion of time. This should not be assumed since Descartes explicitly states that this special sense only arises *when* 'time' is distinguished from 'duration,' which seems to imply that time need not be, or is not always, so distinguished.[12] Given the potential confusion arising from the special sense of 'time' that Descartes identifies in *I.57,* I hereby distinguish this sense from a more general notion of time by labeling this special sense as 'time-in-thought.'

Another important interpretive point in analyzing *Principles I.57* is that Descartes uses both the terms 'attribute' and 'mode' in this passage, without distinguishing between them. In the previous chapter, I discussed the fact that Descartes offers a technical distinction between principal attributes and modes in *Principles I.53.*[13] Though Descartes provided this technical distinction just a few pages prior to *I.57,* Descartes does not seem to be employing the distinction when he uses the terms 'attribute' and 'mode' in this text. As Descartes is frequently imprecise in his use of these terms, this fact is neither very surprising nor noteworthy. In *Principles I.55,* for example, Descartes himself claims that the concept that he picks out by 'mode' in this context is "exactly the same as what is elsewhere meant by an attribute or quality."[14] Given that Descartes' is inconsistent in how he uses the terms 'attribute' and 'mode', whenever these terms appear, one must determine their meaning by reference to the context of their usage. Indeed, since *I.57* intends to explain what type of attributes duration and time are (given that some attributes are in things and others are merely in thought), the term 'attribute' here cannot refer to a primary attribute of substance. As Descartes identifies a unique primary attribute for physical things (extension) and minds (thinking), he cannot be suggesting that duration and time are additional primary attributes. Rather, it seems that *I.57* is just using the term 'attribute' to designate a feature of a substance that depends on that substance for its existence, that is, as a way (either essential or nonessential) for a substance to be. According to this usage, it is reasonable for Descartes to use the terms 'attribute' and 'mode' interchangeably, since both terms pick out this dependent relationship in every context.

There is one more, quick interpretive point to note. In the first sentence of the text, Descartes claims that certain attributes or modes of a substance are in the very substance upon which they depend, while others are only in our thought. As an example of this type of distinction Descartes offers his distinction between duration and time-in-thought. From the fact that Descartes' explanation assumes that *some distinction or other* can be drawn between the two attributes, one cannot immediately assume that he is claiming that duration and time-in-thought are two numerically distinct kinds of attributes, since he might just be claiming that duration and time-in-thought are two numerically distinct *ways of conceiving* of one and the same attribute. In other words, this passage does not explicitly indicate whether Descartes is drawing an ontological or a conceptual distinction between duration and time-in-thought. Moreover, pre-reflectively it might seem as though the distinction is merely a conceptual one. Reading the distinction as conceptual could plausibly be supported by focusing on the interesting "when" that Descartes uses to preface his distinction. In claiming that duration and time-in-thought are only distinguished *when* one conceives of time in a particularly way (namely, as "the measure of movement") one might reasonably suppose that the particular way of conceiving of time-in-thought is responsible for the distinction. If the distinction only arose in virtue of a particular way of conceiving, then this would suggest that the relationship between the attributes of duration and time-in-thought could be one of identity—that the only distinction between them arises in virtue of a way of conceiving them. To prove that the distinction between these two attributes is more than just conceptual, one needs to look deeper into *Principles I.57*—in particular into the different ways that motion relates to each.

Time and Motion in Principles I.57

Immediately after Descartes identifies time-in-thought as the measure of movement, he proceeds to relate motion to duration in particular. He claims that "the duration which we understand to be involved in movement is certainly no different from the duration involved in things which do not move." As already noted, this claim offers strong evidence *against* the view that duration (as distinguished from time-in-thought) could bear a very close resemblance to the reductivist's thesis that time is dependent on motion. If one and the same duration obtains whether or not something is in motion, then it cannot be the case that duration depends on motion, or reduces to a type of motion.

Though *I.57* indicates that duration is not dependent on motion, motion's role for time-in-thought is less clear. There seems to be good reason to anticipate that motion will play a more important role in time-in-thought since Descartes identifies time-in-thought as the "measure of motion" and appeals to the celestial motions to explain it. Descartes claims that the durations of the celestial motions are employed so that the durations of all things can be measured, and then he concludes that "we call this duration 'time.'"[15] Though Descartes appeals to the cosmic motions to explain how time-in-thought functions, he is not explicit in explaining how time-in-thought actually relates to these motions.

Though Descartes claims that the cosmic motions are used to measure all durations *via* a comparison, he does not explicitly state where time-in-thought occurs in the comparison process. It seems, for example, that Descartes could be claiming that (i) time-in-thought is the duration of the cosmic motions themselves (i.e., a position akin to celestial reductionism), or (ii) it might be the relation determined in the act of comparing/measuring these motions against other durations (i.e., a view akin to relationalism), or it could be (iii) an idea discovered or employed when comparing these motions. Even though I have not yet established which of these options best explains time-in-thought, one can already see that any of these choices would give motions a more significant role in time-in-thought than it has for duration. Indeed, under any of these three interpretations, it seems that time-in-thought depends on motion in some way or other. Though *I.57* alone does not provide sufficient grounds for ruling out interpretations (i), (ii), or (iii),[16] it does indicate that motion is somehow required for time-in-thought while it is not required for duration. Thus, I.57 does establish a significant difference between these two temporal attributes.

Recall that in *Rule 12* Descartes asserts that the concepts of motion, duration, and time are such that conceiving of one somehow involves conceiving of the other(s). What was not previously determined was whether or not *Rule 12* distinguishes between two different concepts by the terms 'duration' and 'time.' From *I.57* we can see that 'duration' and 'time' (at least occasionally) pick out two distinct attributes. When the terms pick out duration and time-in-thought, Descartes must identify two, really distinct kinds of attributes since duration is *not* dependent on motion, while time-in-thought *is* dependent on motion in one way or another. Since one and the same thing could not be both *in*dependent from and *depend*ent on motion, it must be the case that Descartes posits duration and time-in-thought to be numerically distinct types of attributes.

Accordingly, *Principles I.57* provides a ready solution to the two, apparently contradictory ways that Descartes relates time and motion. Though it's not clear that Descartes intends the *Principles I.57*'s distinction every time he uses the terms 'duration' and 'time,' when he does employ this distinction, he references two numerically distinct temporal attributes. As this numerical diversity suggests a way for him to hold that time is dependent on motion and that motion is dependent on time without thereby implying a contradiction, it seems that whenever Descartes' relates duration, time, and motion one must consider the possibility that he is employing this distinction. By remaining mindful of this possibility, one can avoid significant interpretive puzzles, since Descartes can be understood as referencing two distinct temporal attributes, each of which bears a contrary dependence relation to motion.

Though interpreting Descartes as a temporal dualist explains the type of distinction referenced in *Principles I.57,* and resolves an apparent inconsistency, I don't think that every reference Descartes makes to 'duration' or 'time' assumes this distinction. Indeed, I think that the fact that Descartes does not always reference this distinction explains why Descartes claims that 'time' is called the measure of movement only *when* it is distinguished from duration in general. Why Descartes might sometimes fail to distinguish 'duration' and 'time' seems to relate to the fact that they are both types of time (now conceived as a genus containing two types of attributes). Nonetheless, to get clear on Descartes' account of time, one must determine if a particular usage of one of these terms is one in which Descartes is discussing the broader genus or one of the particular kinds of temporal attributes.

DESCARTES' DISTINCT TEMPORAL ATTRIBUTES

Having established the fact of Descartes' temporal dualism, I can now proceed toward the goal of explaining the distinct natures of each of the temporal attributes and how they are related. Although Descartes does not clarify whether duration or time-in-thought is more fundamental in *I.57,* it is duration that receives more attention in texts beyond *I.57*. Recall that Descartes states at the beginning of *I.57* that "some attributes or modes are in the very things of which they are said to be attributes or modes, while others are only in our thought."[17] In the sentences following, Descartes makes it clear that time-in-thought refers to the temporal attribute which exists "in thought" and that duration is the attribute "in things." *I.57*'s implication that duration is an attribute inhering in things is a view that Descartes states more explicitly elsewhere. Indeed, when Descartes explains the very nature of duration in *Principles I.62,* he does so by indicating it to be an attribute intrinsic to substances. In *I.62,* Descartes explains that "a substance cannot cease to endure without also ceasing to be,"[18] and thus he claims that duration is merely conceptually distinct from substance.[19] Duration is intrinsic to the nature of substance itself because the duration of a substance is its very enduring. As duration just is the enduring of a substance, it follows that if a substance ceased to have duration then it would simply cease to be.

Since duration is an attribute intrinsic to substance itself it seems to be located at a fundamental level of Descartes' ontology. As such, it is a plausible candidate for being the type of time upon which motion depends. In my earlier analysis, I noted that *Principles II.39* shows Descartes claiming that motion is dependent on some sort of temporal succession. Though noting this, I was not then able to identify what type of temporal succession motion requires (i.e., duration, time-in-thought, or the broader genus of time in general). Having now reflected on duration's nature in particular, it is evident that duration offers the precise type of successive basis that would be a prerequisite for motion. Indeed, Descartes explicitly identifies duration with this sort of successiveness in his correspondence with Arnauld (29 July 1648).

> I understand the successive duration of things in motion, and of the motion itself, no differently from that of things that are not in motion; for earlier and later in any duration are known to me by the earlier and later of the successive duration which I detect in my own thought, with which the other things co-exist.[20]

Descartes' correspondence with Arnauld presents an account of duration that consistently ties together the various other passages that have been considered. First, it suggests that motion (either of bodies or—loosely speaking—of thoughts) is required for the *conceiving* of duration, but that it is ontologically irrelevant. *Rule 12* indicated that duration, time, and motion were somehow tied together in such a way that the ideas of each were somehow implied by the others. The Arnauld correspondence explains this remark with regard to the relation between motion and duration. The concepts of duration and motion are tied together because motion makes duration knowable, whereas duration's inherent successiveness makes motion possible. Duration can serve as the temporal basis of motion since duration is intrinsic to the very nature of a substance, and thus is never found separate from a substance that is in motion. Any substance that is in motion, insofar as it exists at all, has successive duration inherent in its nature. Thus, any substance in motion has its own intrinsic duration that can provide the necessary succession that enables the possibility of its motion.

There are additional reasons to suppose that *Principles II.39* is claiming that motion is specifically dependent on the attribute of duration. In *Principles I.57*, when Descartes is explaining the distinction between duration and time-in-thought, it is duration that he identifies as being a feature of all movement. Recall that he explains that "the *duration* which we understand to be *involved in* movement is certainly no different from the *duration* involved in things which do not move."[21] Accordingly, the duration of a motion seems simply to be the successive stages of a substance's enduring, which are the prerequisite for the substance's moving or changing. Though there can be no motion without the possibility of a substance enduring through successive stages, a substance will endure through successive stages (i.e., have inherent, successive duration) whether or not the substance has any motion. This interpretation explains both what type of time (i.e., successive duration) is required for motion, and why duration is not ontologically dependent on motion. Motion may make duration knowable, but there is duration whether or not it is known.

Though an analysis of duration explains the ontological and epistemic links between motion and duration, it is unclear how time-in-thought fits into this otherwise simple and consistent picture. By excluding time-in-thought from the discussion, one might have argued that "time" in Descartes is just the inherent succession of substances (i.e., duration), which is a feature that is both required for motion and likewise made knowable by motion. However, such an approach would miss the fundamental flavor of Descartes' theory. Descartes offers not one—but *two*—different kinds of temporal attributes. Though the simple account from above explains certain fundamental features of Descartes' account of time, the introduction of Descartes' other temporal attribute (i.e., 'time-in-thought') obscures the picture. Unlike Descartes' account of duration, the account picked out by time-in-thought is not explained by reference to the intrinsic structure of a substance, and it seems somehow to be dependent upon motion. Indeed time-in-thought is a much more elusive attribute than duration.

Part of the obscurity surrounding Descartes' views on time-in-thought is that *Principles I.57* defines it *functionally*. Time-in-thought is called the "measure of movement," Descartes explains, when persons want to measure and compare the extent of various durations. He claims that durations can be measured when the extent of a thing's enduring is compared against "the duration of the greatest and most regular motions which give rise to years and days," and this is called "time." Now that Descartes' account of duration has been clarified, the reason why Descartes recognized a need for time-in-thought is quite evident. Since it has been shown that duration just is the inherent succession found in the very enduring of every substance, it follows that all substances have their own individual durations—the particular enduring of any given substance that grounds the possibility of that substance's motions. As all durations are particular to individual substances, this account does not—by itself—suggest a universal sense of time.

Without the existence of some universal temporality, Descartes' ontology would not admit of any meaningful way of comparing various durations, or even of answering the basic question, "What time is it?" As all durations are particular, if Descartes stipulated no temporal standard that could serve as the measure for all durations, then durations could only be compared on individual levels, according to the only weakly informative measure of faster or slower. If, for example, one were to compare the duration of finger taps on a desk against the duration of drips from a faucet, one could judge that there are four finger taps per single water drip. Though this would indicate a relationship between finger taps and drips, there would be no broader context according to which the taps or drips could be judged. One might conclude that there is a swifter duration between taps than drips (that is, there are more finger taps per drip), but the same judgment would be drawn in a comparison between finger taps and visits from Halley's Comet. Though a comparison could determine that there are many finger taps between each return of Halley's Comet, judging that the duration of successive finger taps is shorter than either faucet drips or visits from Halley's Comet is not terribly informative. Though returns of Halley's Comet and faucet drips may share the slower than finger taps relation, this would provide scant means for organizing the world since no way of relating this information to things that have not been actively compared—like the durations of cat yawns, or drum beats, or camera clicks, and so forth, would thereby be determined. Rather than suggest that the world admits of only very incomplete and partial temporal organization, Descartes appeals to the regular cosmic motions as providing a standard against which all other durations can be compared. In doing so, Descartes makes his universal measure dependent upon motion in a way that duration is not.

Though Descartes' reasons for appealing to the duration of the celestial motions to provide a standard measure are fairly straightforward, the nature of time-in-thought is not. One might have thought that time-in-thought has just been precisely defined as the duration of the regular cosmic motions insofar as they are employed to measure and compare all durations. Though such a simple account is appealing it does not cohere with Descartes' clear contention that time-in-thought is "simply a mode of thought."[22] If the attribute time-in-thought is distinguished from duration in general because it is simply a mode of thought, then it cannot be the case that time-in-thought refers to some extra-mental fact like the particular duration of the regular, cosmic motions. If time-in-thought is a "mode of mind," then it must pick out something found in minds, not celestial bodies. As all modes of minds refer to the principal attribute of a thinking substance, time-in-thought must refer to some particular idea or act of thinking.

If time-in-thought must refer to an idea or act of thought, then one might suppose that time-in-thought does not refer to the particular cosmic durations themselves, but rather to the idea that one extracts from this duration, or to the mental act of comparing this idea against the ideas one has of other particular durations. Though either of these possibilities may seem plausible insofar as they both are consistent with the "mode of thought" language, there are significant puzzles related to either alternative. First, there is the puzzle of determining *which* of these two apparent possibilities actually picks out the nature of time-in-thought. Second, there is the puzzle that either possibility supposes that time-in-thought does not pick out *one* thing. If time-in-thought refers to an idea that individuals are able to extract by observing cosmic durations, then it seems that there would be as many time-in-thoughts as there are minds forming these ideas. The same is true if time-in-thought refers to the act of comparing an idea of cosmic durations against ideas of other durations. According to either interpretation, there is no singular sense of time-in-thought. Instead, either interpretation supposes as many time-in-thoughts as there are minds conceiving or applying the idea of cosmic durations. As such, it is not clear how this multiplicity of time-in-thoughts could provide a singular, universal measure. As a way to begin clarifying these puzzles, I will briefly recall the temporal dualism of Francisco Suarez. As Suarez was a close contemporary of Descartes, if Suarez's account shares significant, common elements with that of Descartes, then perhaps Suarez can help clarify Descartes' account.

DESCARTES AND SUAREZ

In chapter 1, I presented Suarez as offering a Dualist account of time. I there explained that Suarez argues for an account which posited two different kinds of time—intrinsic and extrinsic—with extrinsic time acting as the measure for intrinsic time. The first chapter additionally indicated that Suarez's account was one with which Descartes was likely to be familiar since the view is presented in the *Disputationes* and Descartes makes explicit reference to the *Disputations* in his *Fourth Set of Replies*."[23] In sketching Suarez's account, the first chapter indicated how Suarez's two types of times were not given equal ontological standing. Rather, Suarez identified his fundamental account of time with the inherent enduring of things—what he would frequently refer to as a thing's "intrinsic duration." This intrinsic duration was described by Suarez as being "not distinguished in a thing from the act of being of which it is the duration."[24] Accordingly, Suarez's intrinsic time is clearly very similar to Descartes' account of duration. They both agree that duration is only conceptually distinct from substance. Moreover, just as Descartes credited motion with making duration knowable, Suarez credits motion with being the thing that makes his intrinsic time able to be conceived numerically.

It's been noted that Suarez's intrinsic time, that is, duration, is his fundamental temporal notion. As only conceptually distinct from substance,[25] Suarez locates intrinsic time at the ground level of his ontology. By contrast, Suarez's "extrinsic time" is given a more derivative ontological status. As Suarez's intrinsic duration is particular to enduring things, he posits extrinsic or "imaginary" time as offering an objective standard according to which these intrinsic times can be compared by one's conceiving of these durations as filling up more or less of the imaginary time.[26]

Descartes' time-in-thought bears a marked resemblance to Suarez's imaginary time. Not only does Descartes appeal to time-in-thought in the same way that Suarez appeals to imaginary time (i.e., to explain how individual durations are measured and compared), but Descartes' time-in-thought and Suarez's imaginary/extrinsic time also both appeal to the cosmic motions to explain how this comparison is accomplished. Suarez argues that the imaginary time is able to provide an objective measure *via* an idea one originally forms of the intrinsic duration of particular cosmic motions.

> We conclude that time is unique in the universe because it has its own meaning of extrinsic measurement, viz., in the movement of the heavens...Understood as an extrinsic measurement and as common to the other movements, it is taken partly from nature as it fundamentally exists, or inchoative (as I call it), and partly from a division through reason and an accommodation of it as a means of measurement.[27]

This passage indicates how Suarez takes extrinsic time to be an imaginary construct. Extrinsic time is a constructed idea insofar as it contains features not contained in the idea of the intrinsic durations of the cosmic motions themselves. Unlike the succession of the actual cosmic motions, the infinite succession supposed in one's idea of extrinsic time is an idea of movement that is "continuous, successive, and non-repetitive."[28] The idea that extrinsic time must be infinitely and non-repetitively extended is a requirement of its functioning as an absolute and objective standard against which *any* other duration can be contained and thereby measured. As the cosmic motions are themselves continuously repetitive, to form the idea of an infinitely non-repetitive succession, the original idea provided by the cosmic motions must be altered by one's reason. This altering of the original idea is what Suarez explains as the idea forming "partly from a division through reason and an accommodation of [the duration of the cosmic motions] as a means of measurement." The cosmic motions are thus able to offer a common measure for the measurement of all intrinsic durations provided that the idea of them is combined with the requirements conceived by reason.

In appealing to the cosmic motions as a standard by which intrinsic durations can be measured, Suarez's account of extrinsic, imaginary time clearly resembles Descartes' time-in-thought. As Descartes identified time-in-thought to be a "mode of mind," time-in-thought seems even to share the mental component given in Suarez's account of extrinsic time's being an imaginary construct. As Suarez's account of intrinsic time was likewise shown to closely correlate with Descartes' account of duration, there seems to be good reason to suppose that Descartes' account of time was inspired by Suarez's account. Even beyond their resemblance, such a view is supported by correspondence between Descartes and Gassendi. In a letter to Descartes, Gassendi clearly identifies Descartes' account of time with the sort of dualist account given by Suarez.

> You mean to distinguish between *external and internal time*. But, if what I was talking about above is external time, what, then, is internal time?

> Let us take me for example. I am something that has already been in existence for fifty years which I know is the same duration as all other men of age. The count of the years that have passed since our birth *is not the addition of as many times fifty years as there are of us quiquagenarians*. It is only fifty years. Neither do I recognize nor do I learn from you how to distinguish in any essential ways, my fifty years of age, which are my internal concrete time, from the fifty years which have lapsed since my birth and which for you are external and abstract.[29]

In this passage, Gassendi is clearly presuming that Descartes holds some version of the Suarezarian dualist account of time. If Descartes' own contemporaries interpreted his account of time in this way, then Descartes may have been knowingly and intentionally following the Suarezarian model.

Having compared the accounts of Descartes and Suarez, there is clearly a striking resemblance between the two. Since Suarez spent more time filling out the details of his account, it seems that Suarez's theory might help inform an understanding of Descartes'. Though it cannot be assumed that Descartes' time-in-thought agrees with all the features of Suarez's extrinsic time, the similarities suggest a means of inquiry into time-in-thought. Specifically, it suggests that one consider if Descartes follows Suarez's presumption that time-in-thought has a weaker ontological status than duration. To determine if this similarity obtains, it is necessary to understand the sort of "mode of mind" time-in-thought is for Descartes. Accordingly, this question will be the focus of the next chapter.

CONCLUSION

This chapter has resulted in several important discoveries. First, a resolution to Descartes' seemingly inconsistent ways of relating time and motion has been determined; namely, it was determined that Descartes can consistently maintain that time is dependent on motion and that motion is dependent on time because he posited two distinct types of temporal attributes (i.e., duration and time-in-thought). The former provides the successive temporal basis required for motion, while the latter depends on motion in some fundamental way yet to be determined. In attempting to clarify the individual nature of each of Descartes' temporal attributes a third discovery was made; namely, that there are important similarities between Descartes' account of time and that of Suarez. In considering Suarez's account, one sees the immediate, historical precedent for Descartes' temporal dualism. Moreover, the significant similarities between Descartes' and Suarez's accounts, suggest that Suarez's extrinsic time might provide a model for conceiving of the elusive nature of time-in-thought. To determine just how closely the two views align, a deeper analysis of Descartes' time-in-thought is needed.

NOTES

1. AT VIIIA 64; CSM I.242.
2. AT X 421; CSM I.45–46.
3. AT VIIIA 27; CSM I.212.
4. Aristotle, *The Complete Works of Aristotle*, ed. Jonathan Barnes (Princeton: Bollingen Series, 1995). See: *Physics* IV.14.223a23: 377.
5. AT VIIIA 27; CSM I.212.
6. AT X 421; CSM I.45-6.
7. The Latin may be of some help here. The term here is '*sive*' which is often used in early modern philosophy to indicate an identity (as in Spinoza.) However, *sive* is not always used in this way (even in Spinoza) and so it would be a mistake to automatically assume this sense.
8. From the previous footnote, the use of '*sive*' might provide some *prima facie* evidence that (b) is a more likely interpretation than (a).
9. AT VIIIA 64; CSM I.242.
10. There is, of course, something odd about wondering about chronological primacy when conceiving of the nature of time itself, but this oddity can just be set aside at this point.
11. AT VIIIA 27; CSM I.212. The Latin is as follows:

Alia autem sunt in rebus ipsis, quarum attributa vel modi esse dicuntur; alia vero in nostra tantum cogitatione. Ita, cum tempus a duration generaliter sumpta distinguimus, dicimusque esse numerum motus, est tantum modus cogitandi; neque enim profecto intelligimus in motu aliam durationem quam in rebus non motis: ut patet ex eo quod, si duo corpora, unum tarde, aliud celeriter per horam moveatur, non plus temporis in uno quam in alio numeremus, etsi multo plus fit motus. Sed ut rerum omnium durationem motuum illorum maximorum, & maxime aequabilium, a quibus fiunt anni & dies; hancque durationem tempus vocamus. Quod proinde nihil, praeter modum cogitandi, duration generaliter sumptae superaddit.

12. The Latin reads, "ita, *cum* tempus a duration generaliter sumpta distinguimus, dicimusque esse numerum motus, est tantum modus cogitandi," emphasis added.
13. I previously explained that 'principal attributes' refers to the "principal property which constitutes [a substance's] essence," and that 'mode' refers to a nonessential way that a substance can be.
14. AT VIIIA 26; CSM I.211.
15. AT VIIIA 27; CSM I.212.
16. One might claim that *I.57* excludes (i), that is, the view that time-in-thought might be identical to the duration of the cosmic motions since Descartes claims (just prior to introducing the role of these motions that "if there are two bodies moving for an hour, one slowly and the other quickly, we do not reckon the amount of time to be greater in the latter case than the former." The fact that this passage indicates that the amount of motion in a body is irrelevant to the question of time, might count as denying any sort of identity relation between time-in-thought and the celestial motions if the claim is taken to apply universally to all bodies—including the celestial bodies. That is to say, if Descartes were claiming that the speed of any motions—including say the motions of the sun—are irrelevant to the question of how much time has passed, then clearly this claim would deny any sort of identity between time-in-thought and celestial motions.

Though Descartes might be making the universal claim that the motions of *all* bodies are irrelevant to the question of the amount of time, it is also possible that he is making the less controversial point that faster and slower movements of two non-celestial bodies indicate nothing about the length of time through which the bodies endure. The latter point would be less controversial than a claim including celestial motions since Descartes himself indicates in *I.57* that he follows the common tendency of appealing to the sun's rotation as a measure of time. Though Descartes does not clarify the precise way that the celestial motions function in his account of time, it seems reasonable to suppose that he would not intend to claim that the speeds of all celestial motions are equally irrelevant to the question of how much time a body might have endured through. The speed of the sun's motion, for example, might be relevant to the question of how much "time" has passed since if the sun could make two complete rotations in a 24-hour span, then it might be subject to debate if anything enduring through the sun's swift cycling would have endured for two days rather than one. Thus, when Descartes claims that the speed of a body's motion is irrelevant to the question of time, it seems that there may be some reason to doubt that the claim should be read as making the universal claim about all bodies—including celestial ones. As there appears to good reason to doubt the universal claim, it seems that interpretation that supposes that time-in-thought might be identical to the duration of the cosmic motions cannot be positively excluded based on Descartes' claim that the speeds of two bodies motions are irrelevant to the amount of time through which they endure.

17. AT VIIIA 27; CSM I.212.
18. AT VIIIA 30; CSM I. 214.
19. AT VIIIA 30; CSM I.214.
20. AT V 223; CSMK 358.
21. AT VIII 27; CSM I.212, emphasis added.
22. AT VIII 27; CSM I.212.
23. *Fourth Set of Replies:* AT VII 235; CSM II.164.
24. *Disp. Meta.*, 50.9.1; Stephen H. Daniel, "Seventeenth-Century Scholastic Treatments of Time," *Journal of the History of Ideas* 42 (1981): 592.
25. *Disp. Meta.*, 48.4.15; Daniel, "Scholastic Treatments of Time," 594.
26. *Disp. Meta.*, 40.9.10; Ibid., 594.
27. *Disp. Meta.*, 50.9.16; Ibid., 596.
28. Ibid., 597.
29. Quoted in Piero Ariotti, "Toward Absolute Time: The Understanding and Refutation of the Aristotelian Conception of Time in the Sixteenth and Seventeenth Centuries," *Annals of Science* 30 (1973): 159.

FOUR

Two Temporal Attributes that are Ontologically on Par

An analysis of Descartes' accounts of time-in-thought, duration, and motion has led to the conclusion that Descartes posits two distinct, temporal attributes. This chapter will offer an analysis of the more elusive second attribute, which I have termed "time-in-thought" and which Descartes identified with the "measure of motion" in *Principles I.57*. Despite the terminology I am using, the second attribute is (of course) not the only temporal attribute found in thought. Intrinsic duration is certainly found in minds,[1] but unlike the first temporal attribute, time-in-thought is unique in being the attribute that is *only* found in minds. Apart from knowing that time-in-thought is a "mode of mind," this latter attribute presents significant interpretive challenges. One must minimally determine what kind of mental mode it is, but since Descartes offers few texts that seem to explicitly address this mode, this is not easily done. Nonetheless the task must be accomplished in order to determine if this second attribute is like Suarez's extrinsic/imaginary time. Or, more specifically, to determine if time-in-thought has a lesser, ontological standing than duration. As a derivative status would suggest that duration is the most important feature of Descartes' account, such a determination would suggest that duration is the appropriate focus of study when considering Descartes' account of time. On the contrary, if time-in-thought is of equal ontological standing, as I here argue, then any study that fails to adequately consider time-in-thought is thereby significantly lacking.

In analyzing time-in-thought, I conclude that Descartes' time-in-thought is on an ontological par with duration. From this conclusion I contend that Descartes does not simply offer a type of Suarezarian temporal dualism; rather, he offers his own distinct type of temporal dualism—one that is unique in virtue of its making both temporal attributes equally significant. This point is significant in and of itself and because current scholarship is virtually silent on the topic of time-in-thought in Descartes.[2] Thus, in arguing that time-in-thought and duration have equal ontological standing, I argue that most current understanding of Cartesian time is markedly incomplete.

I claim that time-in-thought possesses equal ontological standing to duration by arguing that time-in-thought is an innate idea, and thus an idea of a Cartesian "true and immutable nature." My argument for time-in-thought's being an innate idea begins with an explanation of Descartes' account of measure in order to clarify what Descartes means when he claims that time-in-thought is a "measure of motion." I then argue that time-in-thought must either pick out an *idea* or an *activity* of minds, before concluding that time-in-thought is an idea (and not an activity). In particular, I propose that time-in-thought picks out an idea like one's idea of the number line, which is employed in measuring, but is not identical to an act of measuring. I establish that this idea is *innate* by i) showing philosophical reasons why Descartes *should think* it is innate, and then ii) compelling textual evidence that Descartes *actually did think* it was innate. By arguing that time-in-thought is an innate idea, I conclude that it is an idea of a true and immutable nature. Accordingly, I argue that time-in-thought is an idea found in minds, but that is only weakly mind-dependent since the content of this idea is not created by minds.

TIME-IN-THOUGHT AS THE "MEASURE OF MOTION"

Though the previous chapter established that time-in-thought is distinct from duration insofar as time-in-thought depends on motion, I have not yet established the nature of time-in-thought's dependence on motion. Indeed, arguing that time-in-thought is dependent upon motion at all is rather odd—given my thesis that time-in-thought is an innate idea.[3] Thus, to argue that time-in-thought is an innate idea, I must explain how time-in-thought can both be dependent on motion and nonetheless be innate. To do this, I must carefully examine the passages where Descartes discusses his concept of measurement.

In *Rule 14*, Descartes offers an extended discussion on the meaning of measure and its role in comparisons. In this rule, Descartes claims that one ought to think of all knowledge (besides simple, pure intuitions) "as resulting from a comparison between two or more things."[4] As such, he asserts that it is the business of reason to prepare for this operation.[5] Such preparation, Descartes explains, begins with one's finding a "common nature" which is participated in by all the things that are compared.[6] These "common natures" become the unit according to which various things can be compared. Descartes explains that these common natures may either be drawn from the things themselves, or can be arbitrarily chosen and applied to the things compared. He writes:

> If no determinate unit is specified in the problem, we may adopt as unit either one of the magnitudes already given *or any other magnitude,* and this will be the common measure of all the others. We shall regard it as having as many dimensions as the extreme terms which are to be compared.[7]

By either identifying or stipulating a common unit among things to be compared, one is able to first measure and then to relate one thing to another. As such comparisons are the source of all knowledge beyond (pure intuitions), identifying a common nature is the key by which most things are known.

Descartes explains that measurement makes a comparison possible by translating various wholes into quantities with identical units—the magnitudes of which can then be compared. According to Descartes, measuring just is the translating of a whole into a collection of parts, since measuring is the converse of counting:

> If we consider the order of the parts in relation to the whole, we are then said to be counting; if on the other hand we regard the whole as being divided up into parts, we are measuring it. For example, we measure centuries in terms of years, days, hours, minutes; if on the other hand we count minutes, hours, days and years, we end up with centuries. It is clear from this that there can be countless different dimensions within the same subject, that these add absolutely nothing to the things which possess them, and that they are understood in the same way whether they have a real basis in the objects themselves or are arbitrary inventions of our mind.[8]

By 'dimension,' Descartes explains that he is picking out "a mode or aspect in respect of which some subject is considered to be measurable."[9] Thus, in the passage above, Descartes asserts that there are as many different "dimensions" for any given magnitude as there are units that one can specify for dividing up the magnitude. It is in this way that measuring is the converse of counting. Whereas in counting one builds up from parts to a whole, in measuring one divides (in thought) a whole into parts.

In *Principles I.57,* when Descartes explains that time-in-thought is the "measure of motion" he claims that the cosmic motions provide the means of measuring all durations. He claims, "in order to measure the duration of all things, we compare their duration with the duration of the greatest and most regular motions which give rise to years and days, and we call this duration 'time.'"[10] Given Descartes' understanding of measurement, he is evidently proposing that the cosmic motions provide the *unit* according to which the durations of all things can be quantified and compared.[11] This unit seems to be something real within the celestial motions, since *Rule 14* explicitly states that "the division of a century into years and days" is something real whereas "the division of the day into hours and minutes is not."[12] The difference in these cases would seem to be that there is a characteristic motion that marks off each day, whereas there is no specifiable motion that relates to hours or minutes. The division of a day into hours or minutes is determined according to the arbitrarily chosen number according to which the day's motion is divided. If a day had been divided by 20 instead of 24, then the length of an hour would be different. The length of a day is, however, determined by the celestial motions and not by a chosen unit.

As the celestial bodies have a specifying motion according to which the unit of a day corresponds, it seems that a day is an identifiable stage in the intrinsic and successive duration of the celestial bodies. Recall that in my earlier analysis of duration (in chapter 3) I determined that each substance has an intrinsic duration, and that this duration is made knowable *via* the motions of a substance. Thus, in the case of the celestial motions, the inherent duration (successive enduring) of the celestial bodies grounds the possibility of their motions, but the regular motions themselves make this successive enduring measurable by specifying knowable, divisible units.

Principle I.57's contention that it is possible to measure the duration of all things *via* a comparison with the regular, celestial motions now seems easily explained. Though the movements of the celestial bodies only specify the intrinsic duration of these particular bodies, the units determinable by these motions can be applied to an analysis of the duration of all things. When the unit of a day is extracted from these motions and applied to an analysis of other durations, this unit provides a "common nature" according to which other durations can be compared. These day units are nothing real in relation to any other substance besides the celestial bodies, but can nonetheless serve as a unit to measure all substances. In other words, when the durations of the celestial bodies are used to measure any other duration, the day unit employed is an example of one of the "adopted" units that Descartes admits as applicable to the problem of comparison. The celestial motions are adopted because they are regular and easily observed.

Though Descartes' account of measure clarifies how Descartes thinks that all durations can be compared, how this comparison relates to his account of time-in-thought is not yet clear. Descartes claims that when the durations of the celestial motions are compared to all other durations that this results in (or perhaps is) 'time' (i.e., time-in-thought). Despite having clarified the nature of this comparison, we have little insight into the nature of time-in-thought itself, because it is not obvious where time-in-thought is found in this comparison process. All that has been discovered is that time-in-thought is part of the process of using the duration of the celestial motions as a unit to compare durations. From this fact alone,

one might think that time-in-thought could pick out the extramental fact corresponding to the number of cosmic motions that have occurred simultaneously with another substance's continued enduring. Such a conclusion would make Descartes' account much closer to the celestial reductionists discussed in chapter 1. However, since Descartes explains that time-in-thought is a mode of thought,[13] it cannot pick out an extramental feature of the world (e.g., some particular amount of cosmic motion). As a mode of thought, it seems that time-in-thought must relate to the mind's comparison of various durations, and not to the extramental facts discovered as a result of such comparisons.

If time-in-thought is a mode of thought, then it seems that it must refer either to the *act* of mind by which various durations are measured, or to some *idea* according to which durations are measured. That is to say, time-in-thought must be either the compar*ing*/measur*ing* of various durations or it is some idea according to which such mental acts are accomplished. Of these two alternatives, it seems that the latter is much more plausible. The former seems less likely for various reasons—not excluding the fact that Descartes calls this mode of mind the measure, and not the measur*ing* of duration. It is unlikely that time-in-thought could refer to an activity of mind because such an activity account would make this mode both strongly mind-dependent and particular. That is to say, if time-in-thought referred to an activity of measuring, then time would not be a common framework within which all substance's endurings are measured. Rather, there would only be individual instances of tim*ings*. If time-in-thought referred to the act of comparing the enduring of substances against the cosmic motions, then there would be numerous time-in-thoughts, not one time-in-thought. Each act of comparing would be its own time. And as noted in the last chapter, part of the reason Descartes wants to introduce time-in-thought is to provide a single, all-encompassing time by which to measure all of the individual attributes of duration. Making time-in-thought into an activity of minds would defeat this purpose.

If time-in-thought is not an activity of mind but instead is an idea in accordance with which durations are measured, then these problems are avoided. In suggesting that time-in-thought is an idea "according to which durations are measured," I propose that time-in-thought functions like the idea one has of the number line. All measuring employs some idea of a number line. How this idea is employed varies according to the thing measured and thus according to the unit chosen for measuring that thing. In the case of spatial measurement, for example, a ruler is employed for measuring length. A ruler is just one's idea of a number line made applicable for spatial measurements *via* the determination of a unit (say, an inch). By positing that time-in-thought picks out an idea like one's idea of the number line, I suggest that time-in-thought picks out an idea that one has *prior* to the determination of a unit. In the case of temporal measuring, *time-in-thought is an idea of an infinitely extended series of successively related parts that is used to measure various durations via the stipulation of a particular unit—namely, the unit given by the movements of celestial bodies.* According to this interpretation, time-in-thought refers to an idea that is integral to the nature of measuring, but is not identical to the act of measuring itself.

WHY TIME-IN-THOUGHT MUST BE AN INNATE IDEA

Having argued that time-in-thought is a kind of idea, I can now consider *what kind* (i.e., fictitious, adventitious, or innate). An initially plausible interpretation might be that time-in-thought is a (fairly uninteresting) adventitious idea that we gain by observing the motion of celestial bodies (much like Suarez's external time.) However, this interpretation is incorrect. While the chosen unit of measurement is certainly based on convention and empirical input, the idea of time-in-thought to which the unit is applied is not. To defend this claim, I will briefly present Descartes' three different categories of ideas, since it is through considering these ideas that one learns which type *could* be consistent with time-in-thought. I can then move from showing why time-in-thought *should* be classified as an innate idea to the texts that indicate that Descartes himself assumed this classification.

Descartes' most famous discussion of his three types of ideas is undoubtedly found in *Med III's* "Trademark Argument."

> Among my ideas, some appear to be innate, some to be adventitious, and others to have been invented by me. My understanding of what a thing is, what truth is, and what thought is, seems to derive simply from my own nature. But my hearing a noise, as I do now, or seeing the sun, or feeling the fire, comes from things located outside me, or so I have hitherto judged. Lastly, sirens, hippogriffs and the like are my own invention.[14]

In this discussion, his account of innate ideas is the most obscure, whereas the other two are more accessible. Adventitious ideas are simply those ideas that one receives through input from the senses upon encountering something in the external world. As Descartes' example indicates, one receives the idea of heat when approaching a fire. This is an adventitious idea because the idea forms independently of one's will as a result of an external cause. By contrast, Descartes describes invented (i.e., fictitious) ideas as being those ideas that one's mind willfully forms. For example, one first has an adventitious idea of a horse and of a horn, and then one willfully unites these two ideas to create the idea of a unicorn. Unlike either adventitious or fictitious ideas, Descartes describes innate ideas as being those ideas that "derive from my own nature." By this Descartes means that innate ideas are not caused by the influence of any part of the created world. Innate ideas are found in created minds *via* the causal efficacy of the creator alone.

In the concluding sections of *Med III,* Descartes describes the divine causation of innate ideas specifically as it relates to his idea of God. He wonders how he might have acquired his idea of God and argues that he could not have gotten it through the senses since it never came upon him unexpectedly. He likewise argues that it could not have been formed by his own invention, since he is "unable either to take away anything from it or to add anything to it."[15] He thus concludes that it must be an innate idea placed in him by God. This discussion is not only explicit in attributing the source of innate ideas to a divine cause; it also begins to suggest a criterion for distinguishing innate ideas from those that are either adventitious or fictitious. Adventitious ideas can be recognized as those ideas that one acquires unexpectedly, and fictitious ideas are those that one is free to alter. Though innate ideas might simply be recognized as being those ideas that meet neither of the above criteria, Descartes provides an additional means for identifying an innate idea as such.

Descartes elaborates on the unique nature of an innate idea in a letter to Mersenne (16 June 1641).

> I use the word 'idea' to mean everything which can be in our thought, and I distinguish three kinds. Some are adventitious, such as the idea we commonly have of the sun; others are constructed or made up, in which class we can put the idea which the astronomers construct of the sun by their reasoning; and others are innate, such as the idea of God, mind, body, triangle, *and in general all those which represent true, immutable and eternal essences.*[16]

In this letter, Descartes' accounts of fictitious and adventitious ideas accords with the *Med III* account. But his account of innate ideas goes further by explaining that innate ideas are ideas of true, and immutable essences. This agrees with *Med III's* claim that innate ideas relate to one's understanding of a thing's nature, of minds and of truth, but the letter to Mersenne explains *why* innate ideas have this property. Innate ideas are truth-entailing because they are neither created by our minds nor derived from the external world. As either such process results in ideas that are limited by the possibilities of human error, neither fictitious nor adventitious ideas are guaranteed to be truth-entailing. On the contrary, innate ideas are truth-entailing because they are the only ones that are divinely caused.

Insofar as innate ideas were previously identified as being ideas caused by God, it is not surprising to learn that these ideas are ideas of truth-entailing, true, and immutable natures[17] since God is not a deceiver. What is particularly interesting about innate ideas being ideas of true and immutable natures is that this implies a test for determining which of our ideas are innate. One can analyze the content of one's idea. If the content of the idea does not admit of alteration, and presents one with eternal truths, then one's idea must be an idea of a true and immutable nature; hence, this idea must have been caused by God.

Before I apply this test to the idea of time-in-thought, I want to briefly clarify the connection between innate ideas, true and immutable natures, and eternal truths that I am claiming. Eternal truths are intimately connected with ideas about true and immutable natures because eternal truths are truths *about* true and immutable natures. In other words, the truths (which we can think of as propositions) that can be drawn from one's idea of a true and immutable nature are eternal truths. These truths are eternal insofar as they were true of the immutable natures prior to and independently of one's discovery of them (through analyzing our innate ideas). That is to say, they are true even when they are not known. These truths, Descartes explains in the *Sixth Set of Replies*, do not "'depend on the human intellect or on other existing things'; they depend on God alone."[18] Though eternal truths exist whether or not they are known, when they are known, this indicates one's possession of an innate idea. So, to pull it together: *true and immutable natures* are what make *eternal truths* true, and an idea of a true and immutable nature is an *innate idea*.

One might recall the similarities between Suarez's extrinsic time and Descartes' time-in-thought and suppose that time-in-thought is more likely to be an adventitious or fictitious idea than an innate idea since these resemble Suarez's extrinsic time. Suarez described extrinsic time as imaginary because it was an idea that one constructed by beginning with the adventitious idea of the cosmic motions and then altering the idea by supposing that this movement is "continuous, successive, and non-repetitive"[19] Through this mental alteration Suarez contends that the idea formed is able to function as an objective measure for all durations. Given the similarities between Descartes' time-in-thought and Suarez's extrinsic time, it is reasonable to suppose that Descartes was aware of and even influenced by Suarez's account. Nonetheless, I do not think it is reasonable to suppose that Descartes' time-in-thought agrees with Suarez's account of how this idea is caused.

It seems that the idea of time-in-thought could not be caused either by input from the external world or by the willful operation of one's reason. Time-in-thought could not be an adventitious idea unless it were an idea that Descartes acquired (independently of his will) by input from the senses. As time-in-thought's function as the measure of all duration has led to the supposition that the idea is of an infinitely extended series of successively related parts, this does not seem to be a good candidate for an adventitious idea. From observing the motions of individual, successive durations one's senses might gain the adventitious idea of successively related parts. As Suarez noted, however, this adventitious idea would be insufficient for one seeking a measure of *all* durations, because it would be limited by the finite, temporal extension of whatever motion was observed.

Even if time-in-thought could not be an adventitious idea, one might have thought that it could be an idea that reason formulates by altering the idea originally given to the mind through experience. But if time-in-thought were thus a sort of fictitious idea, it would have to have a malleable nature, since a defining characteristic of a fictitious idea is that it can be changed at will according to the dictates of one's imagination. In the idea of time-in-thought that has so far been sketched no such malleability is evident. The idea of an infinitely extended series of successively related parts is the "measuring stick" idea assigned to time-in-thought. If this is a fabricated idea, then it seems that it should be possible to alter this idea at will. Yet such alterations would seem to inhibit time-in-thought from serving its function as a universal, objective measure. If the successive parts of the measuring stick could be, for example, randomly rearranged, disordered, or perhaps spread in a non-equidistant manner, then it would lack the requisite regularity needed for an objective measure. Such a disordered idea could not serve as a measure relating all durations. Thus, it seems time-in-thought cannot be a fictitious idea.

Moreover, the mere fact that the idea of time-in-thought includes the idea *of infinitely* extended succession prevents Descartes from supposing that this is a created idea because Descartes denies that minds are capable of formulating a notion of infinite. In his *Med III* proof for God's existence, it is God's infinite nature that Descartes emphasizes as being the characteristic of his idea that his own mind could not have created. Descartes focuses on his idea of an infinite, supremely perfect substance and asks if there is any way that the idea of such a thing could have originated from his own imagination. In virtue of God's infinity, Descartes concludes that it could not. He claims that "it is true that I have the idea of substance in me in virtue of the fact that I am a substance; but this would not account for my having the idea of an infinite substance, when I am finite, unless this idea proceeded from some substance which really was infinite."[20] Descartes adds that his idea of infinite could not have arisen in the way that his ideas of rest and darkness arose; namely, via negations of his ideas of motion or light. "On the contrary," he claims, "I clearly understand that there is more reality in an infinite substance than in a finite one, and hence that my perception of the infinite, that is God, is in some way prior to my perception of the finite."[21] Though one might rightly note that this discussion concerns an infinite substance and not an infinite series *per se*, it seems that similar reasoning would imply that Descartes could not be the source of his idea of an infinite series. If Descartes cannot manufacture an idea of an infinite substance *via* the negation of his idea of a finite substance, then it seems that Descartes would be equally incapable of forming an idea of an *infinitely extended succession* from the negation of the adventitious idea of some limited duration.

One might object to this argument on the grounds that I have ignored Descartes' distinction between the infinite and the indefinite. Perhaps, it might be claimed, we cannot form the idea of an *infinitely* extended succession, but we can form the idea of an *indefinitely* extended succession. The assumption underlying this objection is that there is a difference in kind (or a qualitative difference) between the infinite and indefinite. This objection may seem well-grounded on the distinction Descartes draws in *Principles I.27*. He claims, "since we are finite, it would be absurd for us to determine anything concerning the infinite; for this would be to attempt to limit it and grasp it." Accordingly, he warns that one ought to avoid asserting of things in which we cannot recognize a limit "that it is infinite, and instead regard it as indefinite." In regarding things in this way, Descartes says that he i) acknowledges our finitude, ii) reserves the term of 'infinite' for God alone and, iii) and acknowledges that in the case of all things but God, "our understanding does not in the same way *positively* tell us that they lack limits in some respect." Only the divine nature is such that it positively suggests itself to be a thing that cannot be limited. All other things are such that we can only "negatively" acknowledge that if they have any limits these limits "cannot be discovered by us."[22]

As *Principles I.27* indicates, Descartes does suggest that in deference to our own finitude, we should reserve the term 'infinite' for God. In the *5th Replies*, however, he explains that this does not mean that other things cannot be conceived of as limitless in a single respect.

> In cases like the extension of imaginary space, or the set of numbers, or the divisibility of the parts of a quantity, there is merely *some respect in which I do not recognize a limit*; so here I use the term 'indefinite' rather than 'infinite,' because these items are not limitless in every respect.[23]

In other words, there does not seem to be a distinction in kind (or a qualitative distinction) between the infinite and the indefinite, but only a quantitative distinction. Things which are properly infinite are things which are perceived as having no limits at all. Things which are indefinite are things which are perceived as having *no limit in some specific respect*. Thus, in claiming that the idea of 'time-in-thought' contains the idea of an infinitely extended series, I suggest that the time-in-thought is an idea of *limitless* temporal extension, as opposed to an idea of something properly infinite which would be an idea of something limitless in every respect. Whereas God is infinite in all respects, time-in-thought is infinite in this one respect only. Though the appropriate Cartesian terminology for this kind of limitlessness is "indefinite," I use the term to identify the fact that this succession must lack any limit. The nature of this limitlessness is of the same kind as divine infinity, except that it is an infinity which only relates to the succession being limitless in its duration (and not in every respect, i.e., God).

As a finite/limited being, it seems that I am no more capable of forming the idea of a limitless succession than I am of forming the idea of something that is limitless in every respect. Insofar as limitlessness is a perfection, and one that I lack, it seems I cannot be the cause of an idea that is limitless in any respect. Descartes suggests a very similar point in the *2nd Set of Replies*.

> In my thought or intellect I can somehow come upon a perfection that is above me; thus I notice that, when I count, I cannot reach a largest number, and hence I recognize that there is something in the process of counting which exceeds my powers…[he thus discovers] that this power is something which I have received not from myself but from some other being which is more perfect than I am.[24]

In other words, whenever one happens upon an idea in which the limits of one's own nature are made evident, Descartes suggests that this discovery suggests an external cause as grounding this idea. Accordingly, for substances with finite duration to possess an idea of an unlimited duration, there must be some external source of this idea. Margaret Wilson helpfully clarifies this point.

> Descartes *does* think his power of generating conceptions of the indefinite cannot be accounted for by his own nature, but requires the existence of something outside of himself.... Even though the number series, for instance, differs from God in not being "unlimited in all respects" … Descartes holds that his power of numbering still requires an external cause.... Thus Descartes seems to hold that merely negative knowledge that something is such that we cannot find its limits (if it has them) is sufficient to generate a causal argument [by the Causal Adequacy Principle] for some external existent. [25]

The idea of time-in-thought must be limitless in this way, or it could not measure *any* possible duration. Thus, it is an idea that we could not fabricate/cause from our own natures. Thus, even if the idea of time-in-thought might (more properly) be described as an 'indefinitely' extended series, this correction does not make it an idea that we could create, that is, a fictitious idea.

Though the arguments against time-in-thought's being either adventitious or fictitious count in favor of time-in-thought's being innate, Descartes offers further support for this conclusion. In the *Third Set of Replies*, Descartes uses his causal adequacy principle to provide an additional criterion according to which innate ideas can be recognized as such. Descartes explains that "elements in our thought which do not resemble external objects manifestly cannot have originated in external objects."[26] In other words, if an idea possesses features not found in any external object, then the ideas which have these features must be innate. This follows insofar as adventitious and fictitious ideas both originate from external objects—adventitious ideas as directly given, and fictitious ideas as modifications on what is directly given.

Unlike anything found in the external world, the idea of time-in-thought represents a measuring stick with *necessarily* related temporal parts. If the "moments" on this stick could change their positions, then it could not provide a consistent measure of durations. Indeed, in the *Fifth Set of Replies* Descartes points to this feature of time-in-thought when he distinguishes an idea of time from an idea of duration. Descartes complains that Gassendi tries to evade his *Med III* argument "by talking of the necessary 'connection' which exists between the divisions of time considered in the abstract" while Descartes claims that he is in fact concerned with "considering the time or duration of the thing which endures."[27] With the distinction between duration and time-in-thought in hand, it seems that Descartes' complaint to Gassendi is that Descartes is discussing intrinsic durations, while Gassendi is discussing time-in-thought. Every constituent of the external world is radically contingent on God's continuous (re)creation for it to remain in existence. There are no necessary relations between the different stages of a thing's history. On the contrary, time-in-thought appears to pick out an idea with *necessarily* related, successive parts. As the radical contingency of everything in the external world indicates that nothing in the external world can share this feature,[28] the necessarily related parts of time-in-thought are elements that are not (and indeed could not be) found in any objects in the external world. Thus, as an idea that includes necessarily related parts, time-in-thought must be innate.

TEXTUAL SUPPORT OF A CARTESIAN, INNATE IDEA OF TIME

At this point I can turn from arguing the reasons that Descartes *should believe* that time-in-thought is an innate idea, to showing the very good grounds for thinking that Descartes actually *did believe* this. By looking through a variety of texts, one finds very compelling grounds for thinking that Descartes knowingly and deliberately maintained the existence of an innate idea of time. Evidence of this can be found in *The World*, in some of his correspondence, and even in the *Meditations*.

One bit of textual support can be drawn from Descartes' discussion of the laws of nature. Descartes claims in both *The World* and in correspondence with Mersenne (15 April 1630) that his laws of nature are known *a priori*. In *The World*, he claims that his three laws of nature follow directly from the eternal truths. Descartes claims that through innate ideas "God himself has taught us that he has arranged all things in number, weight and measure. The knowledge of these truths is so natural to our souls that we cannot but judge them infallible when we conceive them distinctly."[29] As a result of our innate knowledge of these truths, Descartes claims that it is possible to know his laws of nature via *a priori* demonstrations.

Likewise, in his correspondence with Mersenne, Descartes claims that God "laid down these laws in nature just as a king lays down laws in his kingdom. There is no single one that we cannot grasp if our mind turns to consider it. They are all *inborn* in our minds."[30] In claiming that the laws of nature are "infallible" and "inborn" in persons, Descartes implicitly adds further evidence for the innateness of time-in-thought. Minds must possess an innate idea of time in order for the laws of nature to be *a priori* derivable, because these laws themselves presume a temporal order. For example, there is an inherent presumption of time in any account of speed, insofar as speed just is the ratio of distance over time. As the following passage indicates, Descartes believes that there are truths about speed (and thus about time) that can be known through an *a priori* demonstration.

> If one part of matter moves twice as fast as another which is twice as large, we must consider that there is the same quantity of motion in each part; and if one part slows down, we must suppose that some other part of equal size speeds up by the same amount.[31]

Since Descartes maintains that his laws of nature can be known *a priori*, and since these laws presume some idea of time (given claims about speed, etc.), it seems that these laws indicate an innate idea of time. Thus, these laws offer evidence in support of time-in-thought being an innate idea.

Descartes offers more explicit claims about an innate idea of time in other correspondence with Mersenne (16 October 1639) and in a letter to Elizabeth (21 May 1643). In his letter to Mersenne, Descartes is discussing the meaning of 'truth.' He notes that 'truth' in the strict sense, denotes the conformity of thought with its object," but he adds that it is also used to pick out things that "can be the objects of true thoughts."[32] Given the truth-entailing nature of innate ideas, it seems this latter sense is the one Descartes employs when identifying "*true* and immutable natures." These natures are 'true' insofar as the ideas of them can be analyzed to arrive at eternal truths. When Descartes discusses this sort of *true* nature, he claims, "no logical definition can be given which will help anyone to discover its nature. I think the same of many other things which are very simple and are *known naturally*, such as shape, size, motion, place, *time*, and so on."[33] In this very interesting passage, "time" is not only listed among those things that can be objects of true thoughts, but also among those things that are known naturally. Such a description clearly implies an innate idea of time.

Likewise, in his correspondence with Elizabeth, Descartes offers another passage which strongly hints at time-in-thought's being an innate idea. In this correspondence, he describes "certain *primitive notions* which are as it were the patterns on the basis of which we form all our other conceptions." Among the very few notions that he lists he includes 'being, number, duration, etc."[34] Given the terminology of this book, listing 'duration' rather than 'time' among these primitive notions seems confused, but since (as has already been seen) Descartes has a tendency to use the terms 'duration' and 'time' interchangeably, this terminological slip does not prevent us from interpreting the "primitive notion" here described as being his idea of time-in-thought. Indeed, since duration has been identified as the extramental,[35] successive enduring of substances, whereas time-in-thought is the "mode of mind" by which these endurings are conceived, it is certainly reasonable to suppose that the temporal idea Descartes describes as being a "primitive notion" is the idea that this book (consistently) calls time-in-thought.

Though Descartes' correspondence with Mersenne and Elizabeth are both quite indicative, Descartes' allusions to an innate idea of time can also be found in the *Meditations*. In *Med I*, for example, when Descartes is considering the things that he can know whether or not he is dreaming, he describes certain "simple" and "universal things"[36] that are known to be real whether or not the things we seem to perceive are real objects or mere fictions formed in a dream. In listing the class of these simple and universal things, Descartes includes "corporeal nature in general...the shape of extended things...the quantity, or size and number of these things...*the time through which things may endure*...and so on."[37] In such a list one clearly sees Descartes referencing things that he commonly classes as innate ideas—shape (e.g., triangle), number, and so forth. In referencing some sort of time that is known whether or not the things we perceive are mere fabrications of dreams, Descartes indicates a type of time that is the sort of thing "from which we form all the images of things, whether true or false, that occur in our thought."[38] Understanding what kind of "time" Descartes is here referencing is clarified in the French version of this clause. In the French version, when Descartes lists the simple and universal things that he knows whether or not he is dreaming, he refers to "the time which measures their duration."[39] Such phrasing clearly points to the *Principles I.57* distinction, and thus clarifies that the *kind* of time that he is describing as being a thing he has and applies to all experiences (whether real or imagined); namely, he identifies the simple, universal idea as time-in-thought.

Likewise, when looking at the introductory paragraphs of *Med V*, one again finds Descartes referencing an innate idea of time in a very significant text. In the beginning of *Med V*, before Descartes has attempted to prove his epistemic access to the external world, he does an inventory of the knowledge he already can know with certainty given that these things are both clearly and distinctly understood and that God is not a deceiver.

> Quantity, for example, or 'continuous' quantity as the philosophers commonly call it, is something I distinctly imagine. That is, I distinctly imagine the extension of the quantity (or rather of the thing which is quantified) in length, breadth and depth. *I also enumerate various parts and local motions; and to these motions I assign various duration.* Not only are these things very well known and transparent to me when regarded in this general way, but in addition there are countless particular features regarding shape, number, *motion* and so on, which I perceive when I give them my attention. And the truth of these matters is so open and so much in harmony with my nature, that on first discovering them it seems *that I am not so much learning something new as remembering what I knew before; or it seems like noticing for the first time things that were long present within me although I had never turned my mental gaze on them before*...I find within me countless ideas of things which even though they may not exist anywhere outside me still cannot be called nothing; for although in a sense they can be thought of at will, *they are not my invention but have their own true and immutable natures.*[40]

When Descartes describes how he is able to enumerate parts of motions in order to assign durations to motions, it seems very clear what true and immutable nature he is referencing. If there is a true and immutable nature through which Descartes is able to enumerate/measure durations, and this is something that he can be certain of *before* he is certain that an external world exists or is knowable, then the thing that that makes these enumerations possible must be an innate idea. As time-in-thought grounds the measure (and thereby the count) of all durations, if Descartes can "enumerate various parts and local motions; and ... assign [these motions] various durations," then he must possess an innate idea of time-in-thought. Only an innate idea could be known at this point in the *Meditations,* and only an innate idea would have its own true and immutable nature.

OBJECTIONS

Given all of the evidence, I think it is quite clear that time-in-thought is an innate idea for Descartes. Before proceeding with these results, I will briefly pause to address two significant objections that may be raised against this conclusion. First, I will consider whether my claim that time-in-thought is an innate idea conflicts with my claims in chapter 3 that time-in-thought is really distinct from duration insofar as it depends on motion whereas duration does not. Since an innate idea should (it seems) depend on nothing beyond divine causation, this is an objection worth considering. Secondly, I will look carefully at another conception of time-in-thought that seems to suggest the best alternative to my own interpretation; namely, I will argue that it is better to interpret time-in-thought as an innate idea than as a mere abstraction. Though I believe this latter claim is already refuted by the above analysis, I will give it some specific attention since it's a view that may seem to have its own textual support and that has been aptly defended elsewhere.

Against the inconsistency complaint, I concede that innate ideas cannot be dependent on a feature of the physical world. Nonetheless, I believe that I can maintain that time-in-thought is distinguished from duration in virtue of its dependence on motion by noting that this is merely a *functional* dependence. I have repeatedly noted that *Principles I.57* identifies time-in-thought according to its function. Time-in-thought is used to measure and thereby compare all durations. What was also noted is that the idea of a mental measuring stick can serve this measuring function *only* if one specifies a unit according to which all durations can be divided and thereby compared. The unit specified by Descartes is the cosmic motions—these provide the "common natures" which make the "measuring stick" applicable. Thus, time-in-thought, when understood functionally, *is* dependent on motion. Without the unit, the innate idea would not be applicable to the world (although it would still exist.) Thus, I conclude that time-in-thought is an innate idea which does not depend on motion for its *being* but does depend on the motion of the celestial bodies for its *function,* since this motion provides the unit which allows durations to be divided and thereby compared.

Though I am certainly not the first to think hard about the '*tempus*' distinguished from duration in *Principles I.57*, I do believe that I am the first to identify this mode as an innate idea. Among the others who have reflected on this mode, I think there is one best alternative to my view which warrants a closer consideration. I think the best alternative is offered in Geoffrey Gorham's article, "Descartes on Time and Duration." Below is a summary of his view.

> *Tempus* is a mode of thought for Descartes because it is generated by an 'abstraction of the intellect.' In the section of the Principles immediately following the account of time, Descartes declares that universals are modes of thought "in the same way" as time. For example, the universal two is a mere mode of thought because it results "when number is considered only in the abstract or in general, and not in any created things." In this case, if we look at a pair of objects, and direct our attention not to their nature but simply to the fact that there are two of them" we can generate an idea of *two* and later apply it to other pairs. In the same way, time is an abstraction from the concrete duration of things. We attend to some particular feature of the sun, its regular motion, and exclude its size, color, and so on. We assign to this motion a number, *numerus motus* expressed in years or days, and extend this number arbitrarily to the duration of other things. We call this universal measure of duration 'time.'[41]

Gorham's account in this passage seems plausible, and thus, it poses a significant challenge for my interpretation. Though I will indicate some of the good grounds that support Gorham's interpretation, I will show why the weight of evidence still agrees with my view.

In claiming that (what I call) time-in-thought is an abstraction that we "generate," Gorham's account clearly contradicts the view that it exists as an innate idea. Though I have argued that the content of this idea (i.e., that it is an idea of a limitless succession of necessarily related parts) is sufficient for showing that this idea is not one that could be caused *via* abstraction, Gorham does point to good grounds for thinking we might be the source of this idea. Most aptly, he notes that Descartes discusses universals in the *Principles I.58*—immediately after drawing the distinction between duration and the mode of mind by which it is conceived.

> *Principles I.58*: *In the same way,* number, when it is considered simply in the abstract or in general, and not in any created things, is merely a mode of thinking; and the same applies to all the other universals, as we call them.[42]

Not only does Descartes preface this principle with the telling phrase "in the same way," which does indeed suggest that he is identifying time-in-thought as being a universal, he then proceeds in *Principles I.59* to explain how universals arise. This account, as Gorham correctly explained, is an account that shows how we give rise to our universal ideas. For example, to gain the universal idea of two from observing two things "we direct our attention not to their nature but merely to the fact that there are two of them, [and thereby] form the idea of the number which we call 'two.'"[43] Since Descartes appears to identify time-in-thought as being a universal immediately before claiming that universals "arise" through a specific way of directing our attention (in particular by abstracting common features from those things upon which we attend), it seems reasonable to claim that Descartes held time-in-thought to be an abstraction we create as opposed to an innate idea. Not only does this seem like an obvious interpretation of *Principles I.58–I.59*, it is clearly significant that these passages are immediately adjacent to the distinction Descartes draws between duration and time-in-thought in *Principles I.57*.

Given what I have just granted, it is evident why I claim that Gorham's contention that time-in-thought is a mere abstraction is a good alternative to my interpretation. Though it is a good alternative, I maintain that the evidence that time-in-thought is an innate idea is stronger. In arguing this, I don't deny that *Principles I.58–I.59* do suggest solid grounds for supposing that time-in-thought is a universal idea, and that this idea can arise *via* a particular mode of attention. I argue, however, that these facts alone provide no reason for thinking that the idea is not innate.

I concede these points both because of the textual evidence which supports them, and because they simply seem correct. I concede that time-in-thought is a universal idea for Descartes since, on my interpretation, it is a quintessential example of us "mak[ing] use of one and the same idea for thinking of all individual items which resemble each other."[44] In particular, it is the single idea used to conceive of *all* substances which resemble each other in virtue of being things that endure. Of course, since every substance will resemble each other in this way, it follows that it is a single idea used to conceive every substance, and not just some subset of existing substances. Given that it is an idea that we use to conceive this feature that characterizes all substances, time-in-thought is a *most* universal idea. Indeed, it is properly described by Descartes as one of the "primitive notions which are as it were the patterns on the basis of which we form *all* our other conceptions."[45] Recall that in his correspondence with Elizabeth he described such general notions, and claimed: "There are very few such notions," the most general being: "those of *being, number, duration, etc.—which apply to everything we can conceive.*"[46]

As a most general type of universal idea, time-in-thought allows us to conceive a resemblance amongst all substances. Insofar as this is a universal idea, and universal ideas "arise" through a method of abstraction, why do I still maintain that this idea is innate, and not *merely* an abstraction? To understand this, one can look to other useful passages, as well as simply consider what means might be employed in order to direct our minds toward thinking about an idea that one already has innately. First, consider the following two passages that are also from the *Principles* and given shortly before *Principles I.57*. First, in *Principles I.49*, Descartes offers a discussion on eternal truths/common notions. He claims that there are certain "eternal truths which reside within our mind. Such truths are termed common notions or axioms."[47] Though he claims that these notions are "common," and "reside within our mind," he adds in *Principles I.50* that "some of them do not really have an equal claim to be called 'common' among all people since they are not all equally well perceived by everyone."[48] In this discussion, one sees that Descartes grants that it is possible for there to be certain eternal truths in our minds that are not (or at least not well) perceived by all. Accordingly, it is appropriate to wonder how one might be led to perceive (or better perceive) these sorts of truths.

Among the eternal truths/common notions described in *Principles I.49–I.50*, Descartes only lists propositions, for example, "what is done cannot be undone."[49] Elsewhere, however, he describes ideas that can likewise reside within our minds and ground eternal truths. An obvious example of such is the idea of a triangle he describes in *Med V*. He claims of a triangle, "even if perhaps no such figure exists, or has ever existed anywhere outside my thought, there is still a determinate nature or essence, or form of the triangle which is immutable and eternal."[50] Likewise in correspondence with Mersenne, he claims, "I use the word 'idea' to mean everything which can be in our thought, and I distinguish three kinds ...[included among the innate are] the idea of Body, mind, body, *triangle*, and in general all those which represent true, immutable and eternal essences."[51] It is particularly useful to note that Descartes lists triangles as being innate ideas that represent eternal essences, because he also discusses triangles when describing the abstracting efforts that give rise to universal ideas in *Principles I.59*.

> When we see a figure made up of three lines, we form an idea of it which we call the idea of a triangle; and we later make use of it as a universal idea, so as to represent to our mind all the other figures made up of three lines. Moreover, when we notice that some triangles have one right angle, and others do not, we form the universal idea of a right-angled triangle.[52]

Insofar as I.59 shows Descartes describing the method by which we "form an idea" of triangle, it seems that this passage provides the same evidence for thinking that triangles are *mere* abstractions as for thinking that time-in-thought is a *mere* abstraction. Yet, to deny that Descartes thinks that one has an innate idea of triangle is simply false. Accordingly, it seems that when *Principles I.59* explains how universal ideas are formed, this explanation does not entail that the ideas so "formed" are not innate. I would contend, rather, that a universal idea may be an innate idea and that *Principles I.59* simply defines a method for directing one's attention to the idea that may already reside within one's mind. There are many passages one can find, where Descartes suggests that this sort of method might be needed.

> *Letter to Voetius, May 1643*: Notice that all things whose knowledge is said to be naturally implanted in us are not for that reason expressly known by us; they are merely such that we come to know them by the power of our own native intelligence."[53]
>
> *1st Set of Replies*: For even if the nature of the triangle is immutable and eternal, it is still appropriate to ask why there is an idea of it within us.
>
> *3rd Set of Replies*: When we say that an idea is innate in us, we do not mean that it is always before us. This would mean that no idea was innate. We simply mean that we have within ourselves the *faculty of summoning up the idea*."[54]
>
> *5th Set of Replies:* I do not, incidentally, concede that the ideas of [geometrical] figures ever came into our mind via the senses . . . Geometrical figures are composed for the most part of straight lines; yet no part of a line that was really straight could ever affect our senses...Hence, when in our childhood we first happed to see a triangular figure drawn

> on paper, it cannot have been this figure that showed us how we should conceive of the true triangle studied by geometers...but since the *idea of the true triangle was already in us,* and could be conceived by our mind more easily than the more composite figure of the triangle drawn on paper, *when we saw the composite figure we did not apprehend the figure we saw, but rather the true triangle.*"[55]

These four passages provide conclusive evidence that Descartes thought innate ideas are not necessarily ideas of which one is always aware. Indeed, they show him describing how one might come to be aware of such ideas. As such, I conclude that there is no inconsistency between me granting that *Principles I.58* and *I.59* show Descartes identifying time-in-thought to be a universal idea that rises into one's immediate thought by means of an abstraction and my contention that it is an innate idea. Rather than indicating that time-in-thought is not an innate idea, *Principles I.58* and *I.59* show how one might draw one's attention to this idea. Since my account seems perfectly consistent with *Principles I.58* and *I.59,* I contend that the various arguments I offered for time-in-thought being an innate idea still hold. Thus, it is correct to regard time-in-thought as an innate idea rather than as an abstraction that one fabricates.

CONCLUSION

Presuming that I have made a compelling case for time-in-thought's being an innate idea, I will now explain a few of the implications of this conclusion. Most significantly, by establishing that time-in-thought is an innate idea, I have explained how this "mode of mind" can serve as a universal measure for all durations, and why it is ontologically on-par with intrinsic duration. To clarify these points, I will indicate how each of these ends has been met.

Recall that an initial motivation for temporal dualism is to solve the problem of measurement. Since intrinsic duration is particular to an individual substance, it provides no means of comparing/measuring various durations against each other. As such, Descartes needed something non-particular that could be universally applied to all these durations. In the innate idea of time-in-thought, Descartes has this universal measure. As an innate idea, the content of the idea is a true and immutable nature created and guaranteed by the non-deceiving God. This idea is singular whether or not the idea happens to manifest in multiple minds. The true and immutable natures themselves are singular. The representations of these natures (i.e., the innate ideas found in minds) are many. This multiplicity does not inhibit the universal, singular character of time-in-thought. In Cartesian terminology, the content of the ideas (their objective reality) is identical even if the formal reality of the ideas is multiple.

As it is the content which makes it possible for time-in-thought to serve its measuring function, the formal differences are no impediment.

The final point to note is how time-in-thought's being an innate idea proves that the second temporal attribute is on an ontological par with the attribute of duration, and thus why Descartes' temporal dualism is significantly different from that of Suarez.[56] Even though duration and time-in-thought are alike in being attributes, there were historical reasons to expect that these two attributes might not be ontological equals; namely, the significant similarities between Descartes' account and that of Suarez. Given that these similarities suggest that Descartes may have intentionally followed the view of his predecessor, there were strong historical grounds to expect that Descartes might also follow Suarez in thinking that duration is something "real" whereas time-in-thought is just imaginary. On the contrary, by proving that time-in-thought is an innate idea for Descartes, I have shown that duration is not the primary temporal category for Descartes. By making time-in-thought an innate idea, Descartes bucks this historical precedent. Time-in-thought is *not merely formulated* by minds in order to serve a measuring function. Unlike Suarez's extrinsic/imaginary time, time-in-thought cannot be created by one's own mind. Time-in-thought cannot be a fictitious idea because the idea of an infinitely extended, necessarily successive, measuring stick is beyond the causal capacities of persons. Thus, for a person to possess this idea it must be innate. As an innate idea, time-in-thought is a much more significant and interesting idea than Suarez's idea of extrinsic/imaginary time. It is more significant and interesting because this status indicates that our idea of time-in-thought is divinely caused and thus can be analyzed as entailing eternal truths.[57]

As duration and time-in-thought are both attributes, and time-in-thought is the idea of a true and immutable nature, there are no plausible grounds for claiming that duration is Descartes' primary temporal attribute, with time-in-thought serving a mere supporting role. Time-in-thought is not subject to duration since it does not depend on duration for its existence. The true and immutable nature of time-in-thought would exist even if there were no bodies. The motions of celestial bodies only serve to make this immutable nature applicable by providing a convenient unit for comparisons. Nonetheless, like all innate ideas, the truths entailed by the idea would be true regardless of any minds possessing the idea, or even of there being any durations to be compared. Thus, time-in-thought is on an ontological par with duration because both are attributes and neither depends on the other for its existence.

NOTES

1. Descartes explicitly claims (in correspondence with Arnauld) that there is successive duration in his own thoughts (See: AT V 223; CSMK 358). Indeed, as duration is just the enduring of any substance, and since minds are one of the two kinds of Cartesian substances, it unquestionably follows that Descartes would recognize intrinsic duration within the progression of minds.

2. In claiming this, I certainly do not claim that others have not noticed or commented on this mode of mind. Rather, I claim that the discussion of this mode has tended to be brief or underdeveloped in such a way that its potential significance has been largely missed.

3. This is of course odd insofar as an innate idea should depend on nothing other than its divine causation—certainly not on motion, and so forth.

4. AT X 440; CSM I.57.

5. AT X 448; CSM I.62.

6. AT X 449; CSM I.63.

7. AT X 449; CSM I.63, emphasis added.

8. AT X 448; CSM I.62.

9. AT X 447; CSM I.62.

10. AT VIIIA 27; CSM I.212.

11. Although the question of whether the "quantifying" of duration (i.e., the whole) consists of duration being broken up into merely potential or actual parts is relevant to the temporal atomism debates in Descartes, this is a question that I am currently setting aside.

12. AT X 448; CSM I.63.

13. AT VIII 27; CSM I.212.

14. AT VII 38; CSM II.26.

15. AT VII 51; CSM II.35.

16. AT III 383; CSMK.183.

17. One might worry that being an idea of a true and immutable nature is merely a *sufficient* condition for being an innate idea, not a *necessary* one. The wording in the letter to Mersenne, however, tells against this suggestion. Descartes claims in this letter (quoted above) that innate ideas are "*in general all* those which represent true, immutable and eternal essences." Thus, I take being an idea of a true and immutable nature to be a necessary condition for being an innate idea.

18. AT VII 436; CSM II. 294. See also: AT VII 104; CSM II.76.

19. Stephen H. Daniel, "Seventeenth-Century Scholastic Treatments of Time," *Journal of the History of Ideas* 42 (1981): 597.

20. AT VII 45; CSM II.31.

21. AT VII 45; CSM II.31.

22. AT VIIIA 15; CSM I.202.

23. AT VII 113, CSM II.81, emphasis added.

24. AT VII 136; CSM II.100.

25. Margaret Wilson, "Can I be the Cause of My Idea of the World?" in *Essays on Descartes' Meditations*, ed. Amelie Oksenberg Rorty (Los Angeles: University of California Press, 1986), 355.

26. AT VII 188; CSM II.132.

27. AT VII 370; CSM II.255.

28. For a discussion of this point, recall *The Argument from Numerically Distinct Parts* given in chapter 2.

29. AT XI 46; CSM I.97.

30. AT I 145; CSMK23, emphasis added.

31. AT VIIIA; CSM.I.240.

32. AT II 597; CSMK 139.

33. AT II 597; CSMK 139, emphasis added.

34. AT III 665; CSMK 218.

35. In calling it "extramental," I don't mean to imply that minds are not successive, but merely that this is a feature of the world that exists both within minds and without minds, and thus is not merely a mental entity. Thus, it is not a good candidate for being a "primitive notion."
36. AT VII 20; CSM II.14.
37. AT VII 20; CSM II.14, emphasis added.
38. AT VII 20; CSM II.14, emphasis added.
39. CSM II.14ff.
40. AT VII 63-64; CSM II.44, emphasis added.
41. Geoffrey Gorham, "Descartes on Time and Duration," *Early Science and Medicine* 12 (2007): 36. Likewise, Arthur claims, "the idea of generic time or duration-in-general...is an ideal concept, formed by abstraction from the durations of all particular substances, just as space or extension-in-general is an abstract concept formed by abstracting from all the particular extensions of pieces of existing matter." Richard Arthur, "Continuous Creation, Continuous Time: A Refutation of the Alleged Discontinuity of Cartesian Time," *Journal of the History of Philosophy* 25 (1988): 361.
42. AT VIIIA 27; CSM I.212, emphasis added.
43. AT VIIIA 27-8; CSM I.212.
44. AT VIIIA 27; CSM I.212.
45. *To Princess Elizabeth*, 21 May 1643: AT III 665; CSMK 218.
46. *To Princess Elizabeth*, 21 May 1643: AT III 665; CSMK 218.
47. AT VIIIA 24; CSM I.209.
48. AT VIIIA 24; CSM I.209.
49. AT VIIIA 24; CSM I.209.
50. AT VII 64; CSM II.45.
51. AT III 383; CSMK 183–184
52. AT VIIIA 28; CSM I.212.
53. AT III 166; CSMK222.
54. AT V11 189; CSM II.132, emphasis added.
55. AT VII 381-2; CSM II.262, emphasis added.
56. One might wonder why there is debate over how to compare duration and time-in-thought since I have argued that they are both attributes and thus both on the same ontological plane. Though it is true that duration and time-in-thought are both attributes and thus that they are both on a general ontological par, the question of how duration and time-in-thought are related is partially founded (and justified) on the comparison to Suarez and partially on a desire to determine if there is some primary temporal attribute for Descartes.
57. Likewise, it is very interesting since it seems to suggest an interesting precursor to Kant's *Transcendental Aesthetic*.

FIVE

Temporal Dualism as an Elegant Solution

Having completed my arguments for the two conclusions: 1) that Descartes offered a type of temporal dualism, and 2) that his dualism was composed of successive duration and the innate idea of time-in-thought, I will here present some important consequences that follow from these conclusions. In particular, I will show how interpreting Descartes in these ways provides a solution to some of the enduring puzzles that have plagued interpretations of Descartes' account of time. Insofar as these puzzles have dominated the scholarship, by showing how my view addresses these puzzles, I show both how my interpretation relates to the broader scholarship and the particular strengths of my interpretation. Indeed, my interpretation's ability to elegantly resolve a number of significant puzzles in the literature suggests final, compelling support for my overall view.

In this chapter, I argue that the continuity of duration is ontologically prior to its having successive parts, since it is continuous in its creation, but divided into parts as a joint consequence of its being conceived *via* time-in-thought and of God's capacity to accomplish whatever can be clearly and distinctly conceived. The structure of the chapter is as follows: In section 1, I offer a brief sketch of how Descartes' account of time has been commonly interpreted by scholars, while also indicating the difficulties found in these interpretations. In section 2, I indicate how interpreting Descartes as a temporal dualist avoids the pitfalls that hinder the standard interpretations, while also offering other interpretive benefits.

THE SCHOLARLY DEBATES

A quick glance at the secondary literature on Descartes' account of time makes the general character of these debates quite clear. In general, scholars debate whether Descartes offered an account of time that is composed of continuous or discontinuous parts, and whether these parts are discrete, temporal atoms or instantaneously temporal slices. Despite the significant discussion on these questions, very little consensus has been reached (perhaps excluding the consensus that interpreting Descartes' account of time is deeply problematic). The view that Descartes held time to be composed of discontinuous temporal atoms has been called the "received" view, but as there has been significant resistance to this view by those claiming that Cartesian time should instead be understood as continuous,[1] it is now, as Geoffrey Gorham contends, more apt to call it the "traditional" or "classic view."[2]

Why it was traditionally supposed that Descartes offered an account of time that consisted of discontinuous, temporal atoms is not difficult to understand. Indeed, this seems to be the most natural way of interpreting Descartes' descriptions of the "parts of time" and the radical contingency that follows these parts. As many have noted, however, this atomistic interpretation seems inconsistent with other essential tenets of Cartesian metaphysics; most notably, it is difficult to reconcile these discrete bits of duration with Descartes' contention that substances maintain their existence by means of God's continuous and immutable act of sustenance. By considering both the advantages and disadvantages of the traditional, atomistic interpretations, I will isolate a few general requirements of a plausible Cartesian account of time. Accordingly, I will briefly lay out the general arguments that have been offered on both sides of the atomism debates. I will not exhaustively categorize the various nuances in these debates, since a less-refined sketch is sufficiently for conceiving of their general structure. Indeed, I will not attempt to present or refute the position of any particular interpretation, since a broader view will better highlights the key features.

As Descartes offers so few passages that directly describe time, it is not surprising that the passages that support an atomistic interpretation are all ones that we have already considered. All the places where Descartes argues from the parts of time to substances' radical contingency on divine concurrence suggest themselves as evidence for an atomistic account of time. Among the most obvious passages are the following:

> *Med III*: For a lifespan can be divided into *countless parts, each completely independent of the others,* so that it does not follow from the fact that I existed a little while ago that I must exist now, unless there is some cause which as it were creates me afresh at this moment—that is, which preserves me. For it is quite clear to anyone who attentively considers the nature of time that the same power and action are needed to pre-

serve anything at each individual moment of its duration as would be required to create that thing anew if it were not yet in its existence.[3]

> *1st Set of Replies*: Now I regard the *divisions of time as being separable from each other*, so that the fact that I now exist does not imply that I shall continue to exist in a little while unless there is a cause which, as it were, creates me afresh at each moment of time.[4]

> *Principles I.21*: For the nature of time is such that its *parts are not mutually dependent, and never coexist*. Thus, from the fact that we now exist, it does not follow that we shall exist a moment from now, unless there is some cause—the same cause which originally produced us—which continually reproduces us, as it were, that is to say, which keeps us in existence.[5]

The inference from these passages to the conclusion that these parts are temporal atoms is pretty straightforward. The language is clearly suggestive of discrete bits of time. Descartes describes the "divisions of time as being separable from each other" and claims that "the nature of time is such that its parts are not mutually dependent, and never coexist." These phrases seem naturally to suggest an account of time in which it contains parts that are discrete, separable bits—consistent with temporal atoms.

Following the suggestive parts language, it was traditionally accepted that Descartes offered an atomistic account of time. Among atomistic interpreters, however, there was debate over the structure of this atomism;[6] in particular, whether the various temporal atoms should be continuously or discontinuously connected (i.e., whether these atoms should be in direct contact with each other or separated by temporal gaps). This debate followed the causal theory that these same passages suggest. In emphasizing the need for an immediate cause for a thing's continued existence, Descartes indicated a causal supposition that he elsewhere made more explicit; namely, that a cause cannot be temporally removed from its effect.

> The natural light does not establish that the concept of an efficient cause requires that it be prior in time to its effect. On the contrary, the concept of a cause is, strictly speaking, applicable only for as long as the cause is producing its effect, and so it is not prior to it.[7]

That Descartes believed a cause must be immediately present to its effect provides a plausible explanation for why Descartes appeals to the "parts of time" in order to explain the radical contingency of substances. A substance cannot ground one moment of its existence on a former moment, suggests Descartes, because these moments are temporally separated from each other. This separation excludes one from continuously self-causing since a cause must be simultaneous with its act of caus*ing*.[8]

It is, however, not obvious what sort of separation would be required between time atoms for them to not be causally efficacious on each other. Whereas some have thought that the very distinctness of these temporal atoms would be sufficient for guaranteeing that one atom could not cause the next, others have wondered if the atoms needed to be more than just discrete—whether they also needed to be discontinuous, that is, divided by temporal gaps. If the various time atoms were immediately conjoined, suggests this latter (discontinuous) view, then it is possible that one moment of a substance's enduring would be conjoined to the next in a way that would not require an additional divine cause to maintain the substance in existence.

Though discontinuous temporal atoms would certainly guarantee that one moment of a thing could not be the causal ground of the next, and would be consistent with the language of independent parts of time, I tend to agree with most of Richard Arthur's criticisms of this view in "Continuous Creation, Continuous Time." As Arthur notes, "Surely if creation is really discontinuous, then it cannot be equivalent to conservation: repeated creation is, one would think the very opposite of conservation."[9] Moreover, if a thing's lifespan really looked like "a repeated alternation between the state of being and the state of nonbeing,"[10] then this would not only be a very strange type of conservation, but it would contradict Descartes' explicit claims in correspondence with More. In this correspondence, Descartes claims, "I think it involves a contradiction to conceive of any duration intervening between the destruction of an earlier world and the creation of a new one."[11] Thus, for Descartes to assert that God created discontinuous, temporal atoms he would need to assert what he elsewhere takes as contradictory.[12]

Even more damaging to those who interpret Descartes as offering discontinuous, temporal atoms are the voids supposed by such views. For there to be temporal atoms that are separated by gaps, the gaps would need to be types of temporal voids. Such voids are clearly not possible for Descartes. Here again is an argument that many others have seen and noted before me, so I will simply sketch the key points.[13] Descartes is very explicit that there can be no spatial vacuums.

> *Principles II.18:* It is no less contradictory for us to conceive of a mountain without a valley than it is for us to think of the concavity apart from the substance which is extended; for. ...nothingness cannot possess any extension...When there is nothing between two bodies they must necessarily touch each other. And it is a manifest contradiction for them to be apart or to have a distance between them, when the distance in question is nothing; for every distance is a mode of extension, and therefore cannot exist without an extended substance.[14]

An empty bit of extension is a "manifest contradiction" within Cartesian metaphysics, because extension is a property of bodies. Thus, when there is no body, there is no extension. An exactly parallel argument excludes the possibility of an empty bit of duration. Duration is an attribute of substances. It is the very enduring of a substance—"a mode under which we conceive the thing in so far as it continues to exist."[15] Consequently, it is a "manifest contradiction" to suppose that there might be bits of duration apart from a thing that endures. Just as it is impossible for there to be bits of extension apart from a body that is extended, it is impossible for there to be bits of duration apart from a thing that endures. There can be no mode for the continued existence of nothingness, since it is a "common notion that nothingness possesses no attributes."[16]

Though I think that discontinuous, temporal atomism is particularly unlikely, any form of temporal atomism is problematic for Descartes—whether the atoms are conceived as continuous or discontinuous. The very idea that Descartes might have posited the existence of temporal atoms (as many have noted) is odd on its face. Descartes was explicit in his rejection of spatial atoms, and it seems that Descartes' grounds for rejecting spatial atoms would have compelled a similar rejection of temporal atoms.[17] In *Principles II.20,* Descartes discusses why spatial atoms are impossible.

> We also know that is impossible that there should exist atoms, that is, pieces of matter that are by their very nature indivisible...for if there were any atoms, then no matter how small we imagined them to be, they would necessarily have to be extended; and hence we would in our thought divide each of them into two or more smaller parts, and hence recognize their divisibility. For anything we can divide in our thought must, for that very reason, be known to be divisible...Even if we imagine that God has chosen to bring it about that some particle of matter is incapable of being divided into smaller particles, it will still not be correct, strictly speaking, to call this particle indivisible. For, by making it indivisible by any of his creatures, God certainly could not thereby take away his own power of dividing it, since it is quite impossible for him to diminish his own power.[18]

In this discussion, Descartes is only discussing spatial atoms, but it seems as though the same line of reasoning is applicable to the possibility of temporal atoms. Insofar as one conceives of atoms as being discrete bits of duration, these bits have some extension. The extent of the duration, no matter how slight, is always such that one could conceive dividing the extent into smaller parts. If such a division is conceivable, then given the fact that God cannot diminish his own power, the bits are in fact divisible in the way that they are conceived to be. Thus, the only way that temporal parts could avoid violating divine omnipotence would be if the parts lacked any duration, that is, were bare instants.

The bare instant interpretation might solve the divisibility problem, but it does so at the cost of creating its own significant problems. To conceive of divine creation as creating instantaneous, non-durational bits is just as problematic as conceiving of this creation as being of extended indivisibles, because these instants could not function as proper parts.[19] Though Descartes' physics does describe the existence of unextended, points of time[20] these points can only serve as limits, and not as parts, if they truly lack any extension. Where there is no extension, there is only a limit, but a limit cannot exist without that which it limits. It is by itself a mere nothing, and no matter how many nothings one has, they cannot be jointly arranged to build up to an extended something.

Whether one interprets Descartes as claiming that time is composed of parts that have duration but are indivisible, or of being composed of parts that lack duration and are thus indivisible, it seems one's interpretation faces serious pitfalls. Thus, if Descartes offered some sort of temporal atomism, then no matter what form this atomism took (i.e., whether continuous or discontinuous), his account was one that was deeply problematic. When one compounds these problems with the problem of how to relate these atoms (i.e., continuously or discontinuously), the problems simply compound. Just as any form of atomism is problematic (by suggesting that an omnipotent God cannot divide some duration, or that one can add many zeros to arrive at a non-zero quantity), there seems to be no metaphysically respectable way for Descartes to link these problematic atoms. If they are discontinuously linked, then this at least explains why a substance would be radically contingent on divine conservation (given the impossibility of a cause at a temporal distance). Yet, this discontinuity suggests a very strange account of conservation. It thus seems that Descartes' "parts of time" can neither have duration nor lack it, and that they can be neither continuously nor discontinuously related.

Despite the obvious problems with Descartes offering any sort of temporal atomism, there is no denying that Descartes' describes time as having parts. He uses the parts language, and appeals to these parts to explain why one moment of a thing cannot be the cause of the next. As these texts cannot be ignored, it seems (or so many have concluded) that Descartes offers a deeply problematic account of time. If his account requires parts, but there is no metaphysically respectable way for him to include these parts, then one might reasonably suppose that there is simply no way to offer a reputable interpretation of Descartes' account of time.

ENTERING THE FRAY

What the prior section undoubtedly emphasizes is the tangle that seems to entrap all Cartesian accounts of time. Not only does Descartes offer very few explicit statements about time, the few he does offer—when teased out—seem to direct one toward untenable theories. Despite the problems with these theories, they do suggest a general guide for what a plausible Cartesian account of time must offer. A plausible account must explain the language of independent parts, as well as their entailment of radically contingent substances. Likewise, the account should avoid crediting a type of temporal atomism that conflicts with other central tenets of Cartesian metaphysics. With inexcusable hubris, I offer my interpretation as doing all of these things.

In my earlier descriptions, I undoubtedly indicated my sympathies. I tend to be very skeptical of any account that entails a discontinuity in the substances that God sustains. Like Arthur, I strongly suspect that alternating states of being and non-being, would be quite the opposite of conservation. Such sympathies are indeed increasingly common in the scholarship. With very few exceptions,[21] recent commentators have shied away from discontinuous accounts of Cartesian time. However, in accounting for how Descartes would cause substances to have *successive* duration (which I have agreed He does do), there are still those who offer accounts which translate into God *successively causing* the distinct parts of created substances.[22] Though these attempts are consistent with Descartes' parts language and avoid crediting Descartes with a conservation fluxuating between being and non-being, I object to the type of divine activity these accounts suggest.

Successive sustaining events seem intuitively to conflict with divine immutability. Descartes' steadfast commitment to the simplicity and immutability of God is beyond question.[23] As such, it is very difficult to conceive how a "God always acting in the same way and consequently always producing substantially the same effects"[24] might be responsible for the fluxuating being supposed by the discontinuous account, *or* for the successive stages suggested by otherwise continuous accounts.[25] Granted, when Descartes claims that God continuously preserves the world "through an action *identical* with its original act of creation,"[26] there is room to say that divine immutability is maintained in virtue of the successive causal acts all being of the exact same type, that is, of reading "identical" in the sense of precise similarity as opposed to the very same act. This looser sense of identity may be defensible, but it suggests a weakened account of divine immutability. It seems that for a strongly immutable God to conserve substances in their being, their being would need to be continuous *and* the divine act that causes this being would likewise need to be simple and indivisible.

Though considerations of the divine nature compel one to suppose that substances must be caused by one simple and continuous act of God, it is unclear how to reconcile this intuition with the fact that Descartes claims that the duration of a substance is successive. That Descartes does maintain the successiveness of duration I have already conceded. Indeed, in chapter 3, I argued that Descartes followed Suarez in supposing that all substances possessed their own intrinsically successive durations that could serve as the ground of their motion. Accordingly, my interpretation must offer some solution to the apparent conflict between God's conserving substances' successive enduring without Himself engaging in successive acts. This is where time-in-thought assists my account by serving a (somewhat Kantian) role in organizing and structuring duration.[27]

Insofar as I have argued that time-in-thought is an innate idea, that is, a primitive notion according to which we conceive of duration, it seems that this idea can be the ground of the divisions in duration, as opposed to some inconstancy in our mode of creation. As time-in-thought is a sequentially ordered measuring stick, its application to duration divides duration into parts. Recall that in chapter 4 I explained how Descartes' notion of measuring just is the translating of a whole into a collection of *parts*. As Descartes claimed, "If we regard the whole as *being divided up into parts*, we are measuring it."[28] Thus, if time-in-thought is the measure of duration, then every time we conceive of duration by means of time-in-thought, we conceive of duration as "divided up into parts."[29] Time-in-thought thus provides a plausible explanation for why "parts" language gets incorporated into Descartes' discussions on time.

In referencing time-in-thought as being the ground of duration's parts, I am not denying that Descartes granted a genuine successiveness to duration. The "parts" created through time-in-thought are not simply how duration is conceived as opposed to how it really is. Recall that I have argued that time-in-thought is an innate idea, and thus an idea of a true and immutable nature. Thus, when time-in-thought structures our conceiving of duration, we learn more than just a fact about how duration is conceived. We learn something true about the nature of duration, since whatever we clearly and distinctly perceive about the world is something that is true of the world.[30] Descartes sees this entailment as following directly from God's not being a deceiver and from His omnipotence. In holding this view, Descartes does not claim that the things we conceive as distinct must ever actually exist separately. Rather, he simply maintains that our clearly conceiving of things as distinct is sufficient for their being divisible, and truly distinct in their natures. Thus, if time-in-thought allows us to clearly and distinctly divide duration into parts, then it is sufficient for determining that duration *actually* has parts.[31]

This analysis provides an explanation for why Descartes seems to use a language slippage when he described the parts of time. As others have noted,[32] Descartes varied between describing the parts of time as separ*able* as opposed to being actually independent. In the *1st Set of Replies*, Descartes described "the divisions of time as being separ*able* from each other" as opposed to separate.[33] Likewise, in *Med III*, Descartes just says that "a lifespan *can be* divided into countless parts, each completely independent of the others,"[34] whereas in *Principles I.21*, he simply asserts that "the nature of time is such that its parts are not mutually dependent, and never coexist."[35] In comparing these various phrasings, it seems that one might sense a sort of discrepancy in the different descriptions. While *Med*

III and the *Replies* each seem to be discussing potential parts, insofar as time is "separable" and "can be divided," *Principles I.21* appears to simply be stating a fact about the nature of time as it is. According to my interpretation, this apparent discrepancy is explained. The parts of duration are *both* potentially separable and actually distinct insofar as God could actually separate any moment from the previous if he so willed. If God willed to withdraw his conservation at any of the points of division that time-in-thought enables us to conceive, then the duration of any substance would cease at that precise moment. If such withdrawal is not willed, however, then the various moments remain continuous and thus are separable without their actually being separated. Through the idea of time-in-thought, God has endowed us with "the faculty of conceiving" which tells us that any amount of endur*ing* is divisible. Thus, "absolutely speaking" there is a genuine distinction between the parts of duration, and not merely a potential one.

The type of distinction between the parts of duration is a bit less clear. I've used the reasoning that distinguishes one substance from another (real distinction), but real distinctions only exist between substances. I argued, however, that duration is an attribute in chapter 2, while I elsewhere noted that Descartes claims there is only a conceptual distinction between a "substance and its duration."[36] Descartes identifies a conceptual distinction as follows: "A conceptual distinction is a distinction between a substance and some attribute without which the substance is unintelligible; alternatively, it is a distinction between two such attributes of a single substance."[37] It was further noted that the distinction between a substance and its duration is a classic example of this sort of distinction—since you cannot conceive of a substance apart from its enduring. When conceiving of distinct parts of duration, however, one might have supposed that the latter half of Descartes' account of a conceptual distinction would apply, that is, "a distinction between two such attributes of a single substance." However, Descartes further explains that a conceptual distinction between attributes is "recognized by our inability to perceive clearly the idea of one of the two attributes if we separate it from the other."[38] Though we do not clearly and distinctly conceive each subsequent moment of duration *per se*, we do so conceive the various stages of an enduring substance. We do not conceive of each subsequent moment *per se* because these are essentially related (in the ways discussed in chapter 2). That is to say, you cannot recognize some particular moment without determining that moment in relation to the surrounding moments. Yet, the enduring substance itself, which is what would be picked out at any particular moment, is a thing that can be clearly conceived. Thus, there is a real (and not merely conceptual) distinction between these moments.[39]

Insofar as time-in-thought determines a real distinction between the moments of a thing's enduring, it offers significant benefits for my position. The parts that are distinguished *via* time-in-thought are not indivisible atoms that would conflict with Descartes' general rejection of atomism. Indeed, the measuring stick idea of time-in-thought tells us that any bit of duration is divisible at any point. Thus, the view is consistent with Descartes' arguments against atomism. Moreover, this account grants really distinct (non-atomistic) parts to duration without thereby running into the problem of non-durational instants being incapable of "adding up" to an extended duration. This is because duration's continuity is ontologically prior to its really distinct parts. In other words, the whole is prior to its parts; a duration first exists as caused by God, and only has parts in virtue of the divisions that could be made in the duration, which correspond to the sustaining that could be withdrawn by God at a moment determinable *via* time-in-thought.

I am assuming the same sort of continuity offered by Richard Arthur and others.[40] Arthur looks at Descartes' correspondence with Gassendi and notes Descartes' emphasis on individual moments of time seems primarily focused on the radical contingency of substances and not on the nature of time itself. Descartes claims, "We are considering the time or duration of the thing which endures, and here you would not deny that the individual moments can be separated from those immediately preceding and succeeding them, which implies that the thing which endures may cease to be at any given moment."[41] When interpreting this passage, Arthur's analysis (which itself looks to the work of Jean-Marie Beyssade) seems exactly right.

> What Descartes is concerned to refute is not that the durations of things are continuous, but the idea that their continuous existence does not require a cause. It is agreed that an enduring thing has a definite existence at any given moment of its duration. Yet since this existence depends on God's continuous action, and a thing has duration only as long as it exists, there is no moment that is not contingent; without God's action there would be no link between one moment and those that follow it. But provided God acts continuously over an extended time, a body will have a continuous duration."[42]

I agree with Arthur's picture insofar as I think that if an immutable God acts always in the same way, then this would result in substances possessed of continuous durations. My interpretation adds an account of how this constancy and immutability would nonetheless create substances with genuinely distinct parts and contingency at every conceivable moment. These parts are the result of the innate idea of time-in-thought, which provides the method for conceiving of duration. This idea of a sequentially ordered measuring stick is such that we immediately apprehend that a thing so measured could be divided at any point—just as the number line of a measuring stick is divisible at any point. Thus, time-in-thought tells us of our radical contingency at every moment, and of the really distinct parts of our duration.[43]

At this point it seems that one might criticize my account of helping itself to a number of goods, without acknowledging the genuine costs in my view. In particular, it might seem as though I must choose either to claim that divine conservation is perfectly immutable in its mode of creation, and thus that there is no genuine successiveness (perhaps just potential successiveness) in the nature of duration; or, that I must grant a genuine successiveness to duration and thereby acknowledge that this requires successive acts in the divine cause. I clearly cannot deny a genuine successiveness to duration. Not only is such successiveness clearly described by Descartes, but it is also what I employed in distinguishing duration from time-in-thought in chapter 3.[44] Given these compelling reasons, it might seem as though my view should not concern itself with arguing for God's acting with one identical act of conservation. As, however, I believe that this genuine immutability is required for an accurate account of Cartesian metaphysics, I think my view would be flawed to whatever degree it failed to maintain this important tenet.

I don't believe there is any conflict between my claims that duration is genuinely successive (i.e., composed of the really distinct parts that time-in-thought determines) and that God could cause this duration by one simple act. The real distinction between mind and body provides a useful parallel to illustrate my point. Of course Descartes grants a real distinction between minds and bodies. Insofar as minds and bodies are both substances, they are genuinely distinct from each other. Nonetheless, Descartes does not argue that God must create them *via* distinct acts. Indeed, in the *Principles*, he supposes an opposite case.

> Even if we suppose that God has joined some corporeal substance to such a thinking substance so closely that they cannot be more closely conjoined, thus compounding them into a unity, they nonetheless remain really distinct. For no matter how closely God may have united them, the power which he previously had of separating them, or keeping one in being without the other, is something he could not lay aside.[45]

In this passage, Descartes conceives that God may have created mind/body composites as unities that never exist separately. If these composites were created as unities, then there would be no reason to suspect that they would require separate acts of creation merely on the grounds that they are genuinely distinct substances. One should not be led to infer from the language of "composite" to think that these two substances would need to first exist independently before they might be joined into the composite. Rather, an omnipotent God is capable in one act of creating two distinct substances as a unity that always exist together.

The parallel to the real distinction in minds and bodies provides the defense for my interpretation of Descartes' account of time. A perfectly immutable God, can create perfectly continuous substances, which are nonetheless composed of really distinct parts. Insofar as these parts are really distinct and successively related in the ways that time-in-thought tells us that they are, duration maintains the genuine successiveness it must have to ground motion. Time-in-thought is truth-entailing, and it tells us that duration is extended in such a way that one part may possess characteristics that another lacks, just as minds possess characteristics that bodies lack—whether or not minds and bodies or various moments of duration are always unified. Accordingly, one really distinct part of duration can be in location B, while another can be in location C. Thus, these distinct parts are sufficient for grounding the possibility of motion.[46]

If I am correct in claiming that duration possesses its successive character as a result of conceiving it *via* time-in-thought instead of as a fact about how our durations are caused, then this provides multiple benefits. First, this picture is consistent with the sort of divine immutability whereby "there is always a single identical and perfectly simple act by means of which [God] simultaneously understands, wills and accomplishes everything."[47] Secondly, this explanation protects the eternity of God. In several recent works,[48] Geoffrey Gorham has argued that God's duration must be successive, contrary to what he acknowledges is the more standard view that supposes Descartes follows the Boethian conception of divine eternity as God simultaneously possessing all of his infinite life. Though Gorham bases his argument on several points, a substantial ground for his view is an entailment on how God could be the cause of our successive duration.

> God is so intimately involved with the unfolding of the Cartesian world, it is hard to see how He could be removed from time. Consider, to begin, God's creation of finite souls. Since the parts of my duration are 'completely independent', I will not continue to exist 'unless there is some cause which as it were creates me afresh at this moment – that is conserves me' (AT, 7, 49; CSM, 2, 33). In creating afresh the successive parts of my duration, God's action is always characterized as an ongoing and temporally extended process rather than a 'once and for all' decree.[49]

This discussion shows further support for holding that duration is successive as a result of conceiving it *via* time-in-thought rather than as an effect of our being successively caused. If time-in-thought is the ground of duration's successiveness, then the divine nature is preserved. God's causation can be perfectly immutable and simultaneous, and thus His nature can be the same.

Insofar as my interpretation maintains that Descartes' God is a strictly immutable and atemporally eternal while nonetheless describing how God is able to continuously conserve all substances in a way that leaves them radically dependent on God's causation at every moment of their enduring, my interpretation has significant benefits for Cartesian metaphysics. As the view also avoids the pitfalls of conflicting with Descartes' general rejection of atoms and voids, or of needing to explain how non-durational instants can add up to a thing's possessing an extended duration, it seems that my view offers many advantages over competing theories. Indeed, I contend that the fact that my interpretation is able to address so many of these enduring problems in the scholarship suggests a final and powerful piece of support for the temporal dualism I have defended.

CONCLUSION

As a final conclusion, I would simply note that by focusing on time-in-thought, an attribute that has been noted but perhaps not properly appreciated by others, I am able to offer a new, and potentially valuable interpretation into the scholarship on Descartes' account of time. My account is historically plausible (given its close resemblance to Suarez), historically significant (since it suggests a very interesting precursor to Kant), and of scholarly benefit (since it takes Descartes "parts of time" language seriously without thereby committing him to the troubling entailments of competing interpretations). As such, it seems that this new interpretation has many benefits and a solid ground of support. As such, I propose that Descartes should be read as a temporal dualist, who maintains two kinds of time: 1) intrinsic duration, and 2) the innate idea time-in-thought.

NOTES

1. See, for example: Richard Arthur, "Continuous Creation, Continuous Time: A Refutation of the Alleged Discontinuity of Cartesian Time," *Journal of the History of Philosophy* 25 (1988): 349-75; Bennett, Jonathan. *Learning from Six Philosophers vol. 1* (Oxford: Clarendon Press, 2001), 98; Jean-Marie Beyssade, *La Philosophie première de Descartes* (Paris: Glammarion, 1979); Jean Laporte, *Le Rationisme de Descartes* (Paris: Presses Universitaires de France, 1950).

2. Geoffrey Gorham,"Cartesian Temporal Atomism: A New Defence, A New Refutation," *British Journal for the History of Philosophy* 16 (2008): 626. Similar views are claimed by Arthur, "Continuous Creation, Continuous Time," 350. For a few examples of the traditional view, see: Henri Bergson, *L'evolution creatice* (Paris: Alcan, 1907), 295, 365ff, 370, 373–374; Martial Guéroult, *Descartes' Philosophy Interpreted According to the Order of Reasons: The Soul and God vol. 1*, ed. and trans. Roger Ariew (Minneapolis: University of Minnesota Press, 1984): Chapter 6; Norman Kemp Smith, *Studies in the Cartesian Philosophy* (New York: The Macmillan Co., 1902): 131–132.

3. AT VII 49; CSM II.33, emphasis added.

4. AT VII 109; CSM II.79, emphasis added.

5. AT VIIIA 13; CSM I.200, emphasis added.

6. Such debates are still occasionally appearing in the literature. See for example: Ken Levy, "Is Descartes a Temporal Atomist?" *British Journal for the History of Philosophy* 13 (2005): 627–674; Geoffrey Gorham, "Cartesian Temporal Atomism: A New Defense, A New Refutation." *British Journal for the History of Philosophy* 16 (2008): 625–637. Levy argues for a strongly discontinuous type of atomism and Gorham offers a direct refutation of Levy.

7. AT VII 108, CSM II.78. See also: AT III 428, CSMK.193: "For I admit I am not subtle enough to grasp how something can be acted upon by something else that is not present—which may indeed, be supposed not to exist anymore. Like the whip if it should cease to exist after whipping the top."

8. For a good discussion on this causal demand for a simultaneous cause see: Jorge Secada, *Cartesian Metaphysics: The Late Scholastic Origins of Modern Philosophy* (New York: Cambridge University Press, 2000), 45–72. See also: E.M. Curley, *Descartes Against the Skeptics* (New York: Harvard University Press, 1978), 140; Geoffrey Gorham,"Cartesian Causation: Continuous, Instantaneous, Overdetermined," *Journal of the History of Philosophy* 42 (2004): 396–400; Gorham, "Cartesian Temporal Atomism," 629.

9. Arthur, "Continuous Creation, Continuous Time," 357.

10. As described in Kemp Smith, *Cartesian Philosophy*, 132.

11. AT V 343; CSMK 373.

12. In "Cartesian Temporal Atomism," Geoffrey Gorham points to an additional strong passage which conflicts with discontinuous atomism. He notes that *Principles I.56* describes duration as a thing that "always remains unmodified." (AT VIIIA 26; CSM I 211). Like Gorham, I think that a duration that admitted temporal gaps would be one that was significantly modified.

13. See for example: Arthur, "Continuous Creation, Continuous Time," 357; Martial Guéroult, Martial, *Descartes' Philosophy*, 199; Geoffrey Gorham, "Descartes on Time and Duration," *Early Science and Medicine* 12 (2007): 52; Geoffrey Gorham, "Descartes on God's Relation to Time." *Religious Studies* 44 (2008): 418.

14. AT VIIIA 50, CSM I.230-1.

15. AT VIIIA 26, CSM I.211.

16. AT VIIIA 25, CSMI.210.

17. Though Descartes himself never explicitly discusses temporal atoms in the ways that he does spatial atoms, it is evident that such parallel reasoning applies. Not only is the reasoning relating temporal and spatial extension straightforward, this reasoning was known and explicitly discussed by others prior to Descartes. See Samuel Sambursky, *Physics of the Stoics* (New York: Macmillan Co., 1959): 103.

18. AT VIIIA 51; CSM I.231.

19. This is an idea from Jean-Marie Beyssade, and one that you can find explained by Dan Garber. See, for example, Dan Garber, "How God Causes Motion: Descartes, Divine Sustenance, and Occasionalism," *The Journal of Philosophy* 84 (1987): 571. See also: Kemp Smith, *Cartesian Philosophy*, 131–132.

20. See, for example, Descartes' discussion of the Second Law of Motion. AT VIIIA, CSM I.242.

21. See for example: Levy, "Is Descartes a Temporal Atomist, 627–674. As this position is generally out of favor, however, it is not surprising that one can find a fast and efficient rejection of this account in Gorham, "Cartesian Temporal Atomism," 625–637.

22. See, for example: Gorham, "Descartes on God's Relation to Time," 413–431; and Geoffrey Gorham, "Descartes on Persistence and Temporal Parts," in *Time and Identity*, ed. Joseph Keim Campbell, Michael O'Rourke, and Harry S. Silverstein (Cambridge, MA: The MIT Press, 2010), 165–182.

23. For some of the many places Descartes claims this, see: AT XI 38; CSM I.93, AT XI 43; CSM I.96, AT *Principles I.56*: VIIIA 26; CSM I.211, *Principles II.36*: AT VIIIA 61; CSM I.240, *Principles II.39*: AT VIIIA 64; CSM I.241, *Principles II.42*: AT VIIIA 66; CSM I.243.

24. AT X1.38; CSM I.93.
25. See, for example: Gorham, "Descartes on Persistence," 165–182.
26. AT VIIIA 66; CSM I.243, emphasis added.
27. Having mentioned Kant's *Aesthetic* several times already, I want to briefly sketch the view that I have in mind. In the *Aesthetic*, Kant explains that there are two forms of intuition: space (the form of outer intuition), and time (the form of inner and outer intuition). In explaining how time functions as a form of intuition, Kant explains that time grounds all appearances; that one cannot remove time from appearances because it is a pre-condition for the possibility of having any appearance. As a form of intuition, Kant identifies time as being a feature of our subjective experiences that we bring to the world and through which we access the world, as opposed to an extra-mental fact of the world that we encounter and discover. It is in this way that I see Kant's account as foreshadowed by Descartes' innate idea of time. In arguing that time-in-thought is an innate idea, I argue that it is an idea contained within our inherent, conceptual apparatus. It is an idea that we bring to our experiences of the world (specifically to the enduring of substances), and through the use of it we are able to organize and structure the enduring of the substances we encounter. The two ideas will, of course, diverge in many ways. For example, I think that time-in-thought differs from Kant's form of outer intuition insofar as the application of time-in-thought would tell us about more than merely how duration is conceived—it would tell us about the nature of duration itself.
28. AT X 448; CSM I.62, emphasis added.
29. In support of the plausibility of time-in-thought being the ground of the parts of duration, again recall Suarez. In chapter 1, I explained that Suarez's extrinsic time played a comparable role for intrinsic duration. As I there noted, in *Disp. Meta.*, 50.12.9 Suarez claims, "[Time] is not simply the number of a continuum, but the number of the parts of the same continuum, which in the thing is only number in potency, actualized through the operation and numeration of the mind." In claiming that the parts of a continuum are only potential until they are enumerated by an act of mind, Suarez's account shows a striking resemblance to my interpretation of time-in-thought's role in enumerating the parts of duration. The only difference being that I do not claim that the parts are only *potential* parts.
30. There are many texts that illustrate this point. See, for example, the following three passages. AT VII.78; CSM II.54: *Sixth Meditation*: I know that everything I clearly and distinctly understand is capable of being created by God so as to correspond exactly with my understanding of it. Hence the fact that I can clearly and distinctly understand one thing apart from another is enough to make me certain that the two things are distinct, since they are capable of being separated, at least by God; AT VII.169; CSM II.119: *Second Set of Replies*, Proposition IV: God can bring about whatever we clearly perceive in a way exactly corresponding to our perception of it; AT III.477–478; CSMK 203: To Gibieuf: From the simple fact that I consider the two halves of a part of matter, however small it may be, as two complete substances, whose ideas are not made inadequate by an abstraction of my intellect, I conclude with certainty that they are really divisible. Someone may tell me that though I can conceive them apart, I have no reason to deny their inseparability because I do not know that God has not united or joined them together so tightly that they are entirely inseparable. I would reply that however he may have joined them, I am sure that he can also disjoin them; so that absolutely speaking I have reason to call them divisible, since he has given me the faculty of conceiving them as such.
31. As I here claim that time-in-thought determines that duration has parts, one might worry that there is a tension between this claim and my argument in chapter 3. There I argued that motion depends on successive duration, and not on time-in-thought. But if duration is itself dependent on time-in-thought for its successive parts, then it seems that I must concede that motion is *indirectly* dependent upon time-in-thought. I still think that my analysis in chapter 3 is correct if we restrict that analysis to *direct* dependence. Though motion may be indirectly dependent upon time-in-

thought (insofar as time-in-thought is the ground of duration's successive parts), it is still true that motion is directly dependent on duration. Despite what may ground duration's parts, the parts are genuine features of duration. Thus the parts of duration (which are a real feature of the nature of duration) are the things on which motion depends. I would like to thank an anonymous reviewer for this perceptive insight.

32. See Arthur, "Continuous Creation, Continuous Time," 354; Beyssade, *La Philosophie première de Descartes,* 17; Laporte, *Le Rationisme de Descartes,* 158–159.

33. AT VII 109; CSM II.79.

34. AT VII 49; CSM II.33.

35. AT VIIIA 13; CSM I.200.

36. AT VIIIA 30; CSM I.214.

37. AT VIIIA 30; CSM I.214.

38. AT VIIIA 30; CSM I.214.

39. In arguing for really distinct parts of time, my interpretation seems to suggest the odd account of persistence that is raised in Gorham, "Descartes on Persistence" 165–182, which relates to the grounds Bonnen and Flage used to reject that Descartes offered actual parts in Clarence Bonnen and Daniel E. Flage, "Descartes: The Matter of Time," *International Studies in Philosophy* 32 (2000): 4; namely, that if there are really distinct parts, then these parts are distinct substances. If there are distinct substances at each moment then one "life" would actually consist of a collection of substantial parts. When Gorham presents this view, he accepts that the parts of time doctrine entails that Descartes offers a perdurantist account of persistence. Though I grant a perdurantist account might be an entailment of the view, I don't think that Descartes recognized or intended this entailment. To whatever degree this entailment follows, I think this persistence problem will result from any interpretation that takes Descartes' "parts" language seriously—it is certainly not a problem that is uniquely consequent on my interpretation. Indeed, I suspect that my interpretation weakens the problem by granting genuine continuity to substances by means of their creation. Moreover, I think that the really distinct parts that follow my interpretation are not obviously the sorts of slices that Gorham suggests. Whatever slices exist, on my interpretation, are those that can be conceived *via* time-in-thought. The types of slices that can be conceived, however, are limited by our capacities for conceiving the indefinite divisibility of any kind of extension. Though time-in-thought tells us that our temporal extension can be divided at any point, Descartes' denial of atoms tells us that he does not think that we can conceive of extension being divisible down to final indivisible parts. Likewise, as Arthur (1988) points out, in Descartes' rejection of Galileo's proof that a line is divisible down to an infinite number of parts (373–375), Descartes would likewise seem to reject that a given temporal extension could be divisible down to some instantaneous time slice. Since Descartes would deny that we can conceive of any divisible or indivisible parts, it seems that the really distinct parts formed as a result of what we can conceive are neither atomic nor instantaneously small (i.e., infinitely small), since neither such slices are clearly and distinctly conceived. Any extension is *indefinitely* divisible. Thus, we only clearly and distinctly conceive that there it is divisible at any point, but we do not conceive of actual parts that result from such divisions. Rather, it seems we clearly and distinctly perceive the sorts of parts that Descartes himself describes; namely, the sorts of parts that would obtain if our durations included all of our existence up until the given moment when divine concurrence might be withdrawn. Such an account appears to suggest really distinct parts that better resemble a "growing block" than some unknown number of moment slices.

40. See Arthur, "Continuous Creation, Continuous Time," 349–375.

41. AT VII 70; CSM II 255.

42. Arthur, "Continuous Creation, Continuous Time," 363. What is not evident in this description is if Arthur's account of God would require that God Himself would be temporally extended in the way that Gorham later contends. Gorham, "Descartes on God's Relation to Time," 413–431 and Gorham, "Descartes on Time and Duration," 28–54.

43. The difficulty in this discussion is that I appear to blur an epistemological fact with a metaphysical one. I am not intentionally hedging on whether or not time-in-thought serves a merely epistemic function, that is, to help us *know* that duration is composed of parts, or makes it the case that duration is *actually* composed of parts. There is a blurring of lines here, but this is a feature of the Cartesian system. Insofar as time-in-thought is an innate idea, whatever it clearly and distinctly tells us is the case must actually be the case. Thus, what it allows us to know is also what must be the case *in virtue of what it allows us to know*. In other words, given Descartes' claims that whatever is clearly and distinctly known is actually the case, something that serves this type of epistemic ground also serves as a metaphysical ground sufficient for guaranteeing that whatever is known is *as* it is known to be. Thus, it is true to say that time-in-thought serves both an epistemic and metaphysical function in relation to the parts of duration. Possessing this idea is sufficient for guaranteeing that we can *know* that duration has parts and that duration actually has parts since God can accomplish anything conceivable.

44. As being the feature of duration, which allows duration to be the attribute on which motion *directly* depends.

45. AT VIIIA 29; CSM I.213.

46. I will only briefly acknowledge that this general picture does (in "wild" moments) suggest to me a sort of 4-dimensionalist picture in Descartes. I will neither expand upon this point, nor defend it, since I think would conflict with many other passages in Descartes. Thus, defending the view would be a hefty (and likely unproductive) task. Nonetheless, I acknowledge that my line of reasoning may be consistent with 4-dimensionalism, and since a recent text has convincingly argued that Spinoza offered a 4-dimensionalist account of time, it seems that this point might be worthy of note. See: Jason Waller, *Persistence through Time in Spinoza* (New York: Lexington, 2012).

47. *Principles I.23*: AT VIIIA 14; CSM I.201.

48. See: Gorham, "Descartes on God's Relation to Time," 413–431, and Geoffrey Gorham, "God and the Natural World in the Seventeenth Century: Space, Time, and Causality," *Philosophy Compass* 4 (2009): 862.

49. Gorham, "Descartes on God's Relation to Time," 419.

Bibliography

Annas, Julia. "Aristotle, Number and Time," *The Philosophical Quarterly* 25 (1975): 97–113.

Aquinas, Thomas. *Summa Theologica.* Edited by Anton C. Pegis. Indianapolis: Hackett Publishing, 1945.

Ariew, Roger. *Descartes and the Last Scholastics.* Ithaca: Cornell University Press, 1999.

Ariew, Roger and Alan Gabbey. "The Scholastic Background." In *The Cambridge History of Seventeenth-Century Philosophy,* edited by Daniel Garber and Michael Ayers, 423–453. New York: Cambridge University Press, 1998.

Ariotti, Piero. "Toward Absolute Time: The Understanding and Refutation of the Aristotelian Conception of Time in the Sixteenth and Seventeenth Centuries," *Annals of Science* 30 (1973): 31–50.

———. "Celestial Reductionism of Time: On the Scholastic Conception of Time from Albert the Great and Thomas Aquinas to the End of the 16th Century," *Studio Internazionale di Filosofia* 5 (1972): 91–120.

———. "The Conception of Time in Late Antiquity," *International Philosophical Quarterly* 12 (1972): 526–552.

———. "Toward Absolute Time: Continental Antecedents of the Newtonian Conception of Absolute Time," *Studio Internazionale di Filosofia.* 5 (1973): 141–164.

Aristotle. *The Complete Works of Aristotle.* Edited by Jonathan Barnes. Princeton: Bollingen Series, 1995.

Arthur, Richard. "Continuous Creation, Continuous Time: A Refutation of the Alleged Discontinuity of Cartesian Time." *Journal of the History of Philosophy* 25 (1988): 349–375.

Augustine. *The Confessions of Saint Augustine.* Translated by John K. Ryan. New York: Image Books, 1960.

Bennett, Jonathan. *Learning from Six Philosophers* (Vol. 1). Oxford: Clarendon Press, 2001.

Bergson, Henri. *L'evolution creatice.* Paris: Alcan, 1907.

Bexley, Emmaline. "Quasi-Absolute Time in Francisco Suarez's Metaphysical Disputation." *Intellectual History Review* 22 (2012): 5–22.

Beyssade, Jean-Marie. *La Philosophie première de Descartes.* Paris: Glammarion, 1979.

Bonnen, Clarence and Daniel E. Flage. "Descartes: The Matter of Time." *International Studies in Philosophy* 32 (2000): 1–11.

Broughton, Janet. "Necessity and Physical Laws in Descartes' Philosophy." *Pacific Philosophical Quarterly* 68 (1987): 205–221.

Chappell, Vere. "Descartes' Ontology." *Topoi* 16 (1997): 111–127.

———, ed. *René Descartes, Part I.* New York: Garland Publishing Inc., 1992.

Copleston, Frederick, S.J. 1963. *A History of Philosophy.* Garden City, NY: Doubleday & Company, 1963.

Cottingham, John, ed. *Descartes' Conversation with Burman.* Oxford: Clarendon Press, 1976.

Cottingham, John, and Robert Stoothoff, and Dugald Murdoch. *The Philosophical Writings of Descartes* vols. 1 and 2. Cambridge: Cambridge University Press, 1985.

Cottingham, John, and Robert Stoothoff, and Dugald Murdoch, and Anthony Kenny. *The Philosophical Writings of Descartes: The Correspondence.* vol. 3. Cambridge: Cambridge University Press, 1991.

Crombie, I. M. *An Examination of Plato's Doctrine's Vol. II: Plato on Knowledge and Reality.* New York: Routledge & Kegan Paul, 1963.

Curley, E.M. *Descartes Against the Skeptics*. New York: Harvard University Press, 1978.

Daniel, Stephen H. "Seventeenth-Century Scholastic Treatments of Time." *Journal of the History of Ideas* 42 (1981): 587–606.

Descartes, Rene. *Oeuvres de Descartes*. Edited by Charles Adam and Paul Tannery. Paris: J. Vrin, 1905.

———. *The Philosophical Writings of Descartes Vols 1-2.* Edited and Translated by John Cottingham, Robert Stoothoff, and Dugald Murdoch. Cambridge: Cambridge University Press, 1984.

———. *The Philosophical Writings of Descartes: The Correspondence*. Edited and Translated by John Cottingham, Robert Stoothoff, Dugald Murdoch, and Anthony Kenny. Cambridge: Cambridge University Press, 1991.

Des Chene, Dennis. *Physiologia: Natural Philosophy in Late Aristotelian and Cartesian Thought.* Ithaca: Cornell University Press, 1996.

Diogenes. *Lives of Eminent Philosophers.* Translated by R. D. Hicks. Cambridge: Harvard University Press, 1965.

Duhem, Pierre. *Medieval Cosmology: Theories of Infinity, Place, Time, Void, and the Plurality of Worlds.* Edited and Translated by Roger Ariew. Chicago: The University of Chicago Press, 1985.

Esposito, Constantino. "The Concept of Time in the Metaphysics of Suarez." In *The Medieval Concept of Time: Studies on the Scholastic Debate and its Reception in Early Modern Philosophy*, edited by Pasquale Porro, 383–398. Leiden: Brill, 2001.

Fox, Rory. *Time and Eternity in Mid-Thirteenth-Century Thought.* New York: Oxford University Press, 2006.

Frankfurt, Harry G. "Continuous Creation, Ontological Inertia, and the Discontinuity of Time." In *Necessity Volition and Love*, 55–70. Cambridge: Cambridge University Press, 1998.

———. *Demons, Dreamers, and Madmen*. Indianapolis: Bobbs-Merrill Company, 1970.

Garber, Daniel and Ayers, Michael, eds. *The Cambridge History of Seventeenth-Century Philosophy.* Cambridge: Cambridge University Press, 1998.

Garber, Daniel. "Descartes and Occasionalism." In *Causation in Early Modern Philosophy: Cartesianism, Occasionalism, and Preestablished Harmony*, edited by Steven Nadler. University Park, PA: PSU Press, 1993.

———. *Descartes Embodied: Reading Cartesian Philosophy through Cartesian Science*. Cambridge: Cambridge University Press, 2001.

———. *Descartes' Metaphysical Physics*. Chicago: The University of Chicago Press, 1992.

———. "How God Causes Motion: Descartes, Divine Sustenance, and Occasionalism." *The Journal of Philosophy* 84 (1987): 567–580.

Gassendi, Pierre. *The Selected Works of Pierre Gassendi*. Edited and Translated by Craig B. Brush. New York: Johnson Reprint Corp, 1972.

Gaukroger, Stephen. 1995. *Descartes: An Intellectual Biography.* Oxford: Clarendon Press.

———. *Descartes' System of Natural Philosophy.* United Kingdom: Cambridge University Press, 1995.

Gorham, Geoffrey. "Cartesian Causation: Continuous, Instantaneous, Overdetermined." *Journal of the History of Philosophy* 42 (2004): 389–423.

———. "Cartesian Temporal Atomism: A New Defense, A New Refutation." *British Journal for the History of Philosophy* 16 (2008): 625–37.

———. "Descartes on God's Relation to Time." *Religious Studies* 44 (2008): 413–431.

———. "Descartes on Persistence and Temporal Parts." In *Time and Identity*, edited by Joseph Keim Campbell, Michael O'Rourke, and Harry S. Silverstein, 165-82. Cambridge, MA: The MIT Press, 2010.

———. "Descartes on Time and Duration." *Early Science and Medicine* 12 (2007): 28–54.

———. "God and the Natural World in the Seventeenth Century: Space, Time, and Causality." *Philosophy Compass* 4 (2009): 859–872.

Guéroult, Martial. *Descartes' Philosophy Interpreted According to the Order of Reasons: The Soul and God* (Vol. 1). Edited and Translated by Roger Ariew. Minneapolis: University of Minnesota Press, 1984.
Holden, Thomas. *The Architecture of Matter.* New York: Clarendon Press, 2004.
Huggett, Nick. *Space from Zeno to Einstein: Classic Readings with a Contemporary Commentary.* Cambridge, MA: The MIT Press, 1999.
Joyce, Richard. "Cartesian Memory." *Journal of the History of Philosophy* 35 (1997): 375–393.
Hussey, Edward, ed. *Aristotle's Physics.* New York: Oxford University Press, 1983.
Kaufman, Dan. "Divine Simplicity and the Eternal Truths in Descartes." *British Journal for the History of Philosophy* 11 (2003): 553–579.
———. "God's Immutability and the Necessity of Descartes' Eternal Truths." *Journal of the History of Philosophy* 43 (2005): 1–19.
Kemp Smith, Norman. *Studies in the Cartesian Philosophy.* New York: The Macmillan Co., 1902.
———. *Studies in the Philosophy of Descartes.* London: Macmillan and Co., 1963.
Kenny, Anthony. *Descartes: A Study of his Philosophy.* New York: Random House, 1968.
Kirby, Brian S. "Descartes, Contradiction, and Time." *History of Philosophy Quarterly* 10 (1993): 137–145.
Klein, Julie R. "Memory and the Extension of Thinking in Descartes' *Regulae.*" *International Philosophical Quarterly* 42 (2002): 23–40.
Kretzmann, Norman, ed. *Infinity and Continuity in Ancient and Medieval Thought.* Ithaca: Cornell University Press, 1982.
Laporte, Jean. *Le Rationisme de Descartes.* Paris: Presses Universitaires de France, 1950.
Lauer, Helen. "Descartes' Concept of Number." *Studia Cartesiana* 2 (1981): 137–143.
Levy, Ken. "Is Descartes a Temporal Atomist?" *British Journal for the History of Philosophy* 13 (2005): 627–674.
Lewis, Neil. "Space and Time." From *Cambridge Companion to Duns Scotus.* Edited by Thomas Williams. Cambridge: Cambridge University Press, 2002.
Lolordo, Antonia. *Pierre Gassendi and the Birth of Early Modern Philosophy.* Cambridge: Cambridge University Press, 2007.
Marion, Jean-Luc. *On Descartes' Metaphysical Prism.* Translated by Jeffrey L. Kosky. Chicago: The University of Chicago Press, 1999.
Matthew, Gareth. *Thought's Ego in Augustine and Descartes.* Ithaca and London: Cornell University Press, 1992.
Mooji, J.J.A. *Time and Mind: The History of a Philosophical Problem.* Boston: Brill Academic Pub, 2005.
Moran, Dermot. *The Philosophy of John Scottus Eriugena: A Study of Idealism in the Middle Ages.* New York: Cambridge University Press, 1989.
Nadler, Steven, ed. *Causation in Early Modern Philosophy: Cartesianism, Occasionalism, and Preestablished Harmony.* University Park, PA: PSU Press, 1993.
Nelson, Alan. "Introduction: Descartes' Ontology." *Topoi* 16 (1997): 103–109.
Nolan, Lawrence. "Descartes' Theory of Universals." *Philosophical Studies* 89 (1998): 161–180.
———. "The Ontological Status of Cartesian Natures." *Pacific Philosophical Quarterly* 78 (1997): 169–194.
———. "Reductionism and Nominalism in Descartes' Theory of Attributes." *Topoi* 16 (1997): 129–140.
Pavelich, Andrew. "Descartes' Eternal Truths and Laws of Motion." *The Southern Journal of Philosophy.* 35 (1997): 517–536.
Rau, Catherine. "Theories of Time in Ancient Philosophy," *The Philosophical Review* 62 (1953): 517–518.
Sambursky, Samuel. *The Physical World of Late Antiquity.* Princeton: Princeton University Press, 1962.
———. *Physics of the Stoics.* New York: Macmillan Co., 1959.

Schmaltz, Tad M. *Radical Cartesianism: The French Reception of Descartes.* Cambridge: Cambridge University Press, 2002.

Secada, Jorge. *Cartesian Metaphysics: The Late Scholastic Origins of Modern Philosophy.* New York: Cambridge University Press, 2000.

———. "Descartes on Time and Causality." *The Philosophical Review* 99 (1990): 45–72

Sepper, Dennis L. *Descartes' Imagination.* Berkeley, CA: University of California Press, 1996.

Skirry, Justin. "Descartes' Conceptual Distinction and its Ontological Import." *Journal of the History of Philosophy* 42 (2004): 121–144.

Sorabji, Richard. *Time Creation and the Continuum: Theories in Antiquity and the Early Middle Ages.* Chicago: University of Chicago Press, 2006.

Spinoza, Baruch. *Complete Works.* Translated by Samuel Shirley, Edited by Michael L. Morgan. Indianapolis, IN: Hackett Publishing, 2002.

Suarez, Francisco. *Disputationes Metaphysicae* 50.9-12. In *Opera Omnia,* reprint. Hildesheim: Olms, 1965.

Waller, Jason. *Persistence through Time in Spinoza.* New York: Lexington, 2012.

White, Michael J. *The Continuous and the Discrete.* Oxford: Clarendon Press, 1992.

Whitrow, G. J. *The Natural Philosophy of Time.* Oxford: Clarendon Press, 1980.

Williams, Bernard. *Descartes: The Project of Pure Enquiry.* Brighton: Harvester Press, 1978.

Wilson, Margaret. "Can I be the Cause of My Idea of the World?" In *Essays on Descartes' Meditations*. Edited by Amelie Oksenberg Rorty. Los Angeles: University of California Press, 1986.

Wilson, Margaret. *Descartes.* New York: Routledge, 1978.

Index

About the Author

Rebecca Lloyd Waller is assistant professor of philosophy at Kenyon College. She joined Kenyon in 2009, after completing her PhD at Purdue University. Rebecca's primary research interests are in the history of philosophy and in metaphysics. She has particular interest in the works of Descartes, and in the topics of time and free will. She has published articles on the accounts of time in Leibniz, Spinoza, and McTaggart.

CPSIA information can be obtained at www.ICGtesting.com
Printed in the USA
BVOW01*0124081014

369678BV00003B/5/P

9 780739 175224